Building in the North

Eb Rice

University of Alaska Press
Fairbanks, Alaska

Library of Congress Cataloging-in-Publication Data
Rice, E. F. (Elbert Floyd), 1923–1982.
  Building in the north / by Eb Rice.—5th ed.
      p. cm.
  Includes bibliographical references and index.
  ISBN 978-1-60223-019-4 (pbk. : alk. paper)
  1. Building—Cold weather conditions. 2. Buildings—Environmental engineering.
  3. Buildings—Arctic regions. 4. Architecture and climate. I. Title.
  TH153.R53 2007
  693.8—dc22
                              2007043528

Cover photographs by Jack Schmid: On the cover are infrared and visual spectrum images of the Trans-Alaska Pipeline and a house in downtown Fairbanks.

The bright yellow in the infrared image of the thermosyphons on the pipeline vertical support members shows that the cooling fins are dissipating heat from the subsurface permafrost soils. The cooling fins are at −9°F and the surrounding air is −35°F. This keeps the permafrost cold enough to reduce melting during the warm summer season.

The images of the house, both taken at the same time, demonstrate the capability of the infrared camera to detect heat loss through framing and windows. The air temperature was −25°F and warm parts of the windows were 0°F.

Acknowledgements

This fifth edition of *Building in the North* by Eb Rice is the result of many collaborations, and acknowledgements are in order. Dr. John P. Zarling enhanced and updated several of the graphics and did a thorough technical edit to ensure that the quality of the publication was of the highest engineering standard. John is known for this and his help is deeply appreciated.

Jack Schmid provided several of the photographs on the cover and a technical and cost update for several of the energy cost estimates in the new edition.

The staff of the University of Alaska Foundation have been very supportive in the production of this book. Particularly we would like to thank and acknowledge Mr. Scott Taylor and Kim Davis of the UA Foundation.

Thanks also go to the Institute of Northern Engineering; staff and faculty in this university research center have worked to update and disseminate this treasure of northern engineering since the 1990s, and they were instrumental in bringing together the team that produced this fifth edition. In particular, we thank Douglas J. Goering, Dean of the College of Engineering and Mines, and Sandra Boatwright, Editor for the INE Publications Office, for their support of this project.

Several staff at the Cooperative Extension Service contributed notable technical services to this project, without whom it would never have succeeded. They include Anthea Craven, who did the lion's share of retyping and reformatting the publication for this edition. Amy Simpson and Jeff Fay contributed technical expertise in the production process. They are highly esteemed, competent co-workers and friends.

Sue and Russ Mitchell, always creative and enthusiastic in everything we ask them to do, contributed the design and layout for the cover.

Richard D. Seifert

April 2007

Table of Contents

This fifth edition is dedicated to
Roberta "Bobbie" Rice.

Foreword to the Fifth Edition

In his 1984 foreword to the third edition, Dr. Vince Haneman, then dean of the School of Engineering, notes that Eb Rice will be remembered as long as his students live and as long as the tales of his exploits are repeated. I am proud to be one of those students. Dr. Eb Rice was on my Masters' Thesis committee and he was from the day I met him, a noble teacher who set the standard for professional relationships between students and faculty in an exemplary role as an engineer in practice and in teaching.

There are so many people to thank and many of them are covered in the forewords to the previous editions, including those who have contributed added material to this wonderful document. And as we publish this edition Dr. Rice's wife Roberta has also restored to it some of the original materials from the very first edition published in 1975.

As an admiring student, and as an Extension engineering specialist at the University of Alaska, I wanted to ensure that this enormous legacy of insightful, common sense information enclosed in these covers gets restored to its original value and is continuously available for both students and engineers of the North. Although the world continues to change and much of the change is induced by human enterprise, the wisdom, the insights, and the simple learning from experience that's embodied in this work will serve us all extremely well in the future.

It is with great pleasure and a distinct honor that I, Rich Seifert, now a full professor at the University of Alaska Fairbanks, offer this foreword to the fifth edition of Building in the North by Eb Rice. It is a work of love for the North and may it ever be so.

Richard D. Seifert, Professor
UAF Cooperative Extension Service
Energy and Housing
July 2006

Foreword to the 1996 Edition

This revision of *Building in the North* represents the fourth edition since original publication in 1975. It is based on a series of articles Professor E.F. "Eb" Rice wrote for the University of Alaska Fairbanks magazine, *The Northern Engineer*. When Eb Rice died in 1982, he left a legacy that has been impossible to replicate. During his 30 years on the civil engineering faculty at UAF, he gained a reputation as "Mr. Arctic Engineering" and influenced the lives of thousands of Alaskans and others with his technical expertise and his practical communication skills. His aim for this series, as he wrote in his introduction, was "to provide a basis for the exercise of common sense in design for the North." Our aim in this revision was to remain true to Eb's vision and to provide a brief description of some of the newest arctic engineering techniques.

In his foreword to the third edition on the following page, Dean Emeritus Vincent S. Haneman, Jr., speaks eloquently of the respect Eb Rice commanded: "He was an educator who will be remembered as long as his students live and as long as tales of his exploits are repeated. He was a man I am proud to have known and called friend."

Over the years, as we learned more about building in cold regions, some of the material in these earlier volumes became outdated. Several members of the faculty of engineering at the University of Alaska Fairbanks and Anchorage combined their efforts to review and update the current edition. I served as manager for the project, but most of the credit for rewriting goes to Professors Tom Kinney, Civil Engineering at UAF; Terry McFadden, Mechanical and Arctic Engineering at UAF; John Olofsson, Civil Engineering at UAA; and John Zarling, Mechanical Engineering at UAF. I am grateful for the splendid cooperation, expertise and interest these colleagues have contributed to this project.

We acknowledge financial support from the Alaska Science and Technology Foundation. We appreciate former ASTF Executive Director John Sibert's encouragement of this project and the good advice and support offered by Circuit Rider Kay Slack and others on the ASTF staff.

In the years since Eb's death, his wife, Roberta "Bobbie" Rice has continued her keen interest in cold regions engineering and the engineering programs at the university. With profound appreciation and respect, we dedicate this volume to her.

We have tried to preserve the essence of Eb Rice's original style and content, while providing current information on practice and costs. Such an assignment is delicate at best; we trust that we have succeeded in producing a volume that is both faithful to everything Eb Rice stood for and reflective of current cold regions engineering practice.

F. Lawrence Bennett, P. E.
UAF School of Engineering
November 1995

Foreword to the Third Edition

This is the third edition of a collection of articles by Eb Rice. Eb has left us a real legacy in his compilation of over 20 years of cold climate engineering experience. As Max Brewer said in his original foreword of 1975, "Eb Rice has attempted to bridge the gap between theory and practical, down-to-earth engineering for construction in the Northland. In fact, he does bridge the gap for those serious workers who are not so locked into preconceived notions and temperate climate thinking that they can no longer ask, 'Is there a better way?' Because there is."

The adaptation of temperate zone engineering by minor changes generally does not work without thinking through the fundamentals of arctic problems. Eb has provided the basis for thought in this volume.

Time has passed since the original collection appeared in the *Northern Engineer* and more data have been collected, so additional chapters have been added. His death on April 9, 1982, was a loss to all who knew him, a loss to those who did not, a loss to all in the near and far Arctic, a loss to all not in the Arctic, an irreplaceable loss to us all. Eb's work has so many dimensions of wit, humor, and concern for individuals that it is difficult to conceive of changing his text.

He took great delight in assuming any side of a discussion, even both sides at times, to promote thought. He was an educator who will be remembered as long as his students live and as long as tales of his exploits are repeated. He was a man I am proud to have known and called friend.

The original text was modified in December of 1982 to reflect current costs, data, and practices of the time.

Vincent S. Haneman, Jr., P. E.
University of Alaska Fairbanks
December 1984

From the Author

This selection of articles, originally published in the *Northern Engineer* magazine, was intended to cover some of the special problems that beset the builder of habitations (homes, camps, even warehouses, hangars, stores, or other commercial structures) in cold climates. The articles were assembled under the rubric "The Ideal Arctic House"—but I hope it's clear that the principles governing heat, moisture, frozen ground, insulation, and so on are applicable to all polar facilities, whether or not they are houses.

As the articles turned out, they covered the major principles that make arctic enclosures unique, leaving a miscellany of small topics, each important but unworthy of treatment as a separate article.

So the articles as previously published left out the small things, some of which are not so small, and all of which are worthy of the attention of those who would build in the North.

What Could Go Wrong (Go Wrong...Go Wrong...Go...)? or, Murphy's Law Strikes Again

An old-timer tends to be amazed and exasperated at the number of "errors of judgement" made by newcomers in the North. He often wonders how *anyone* with a brain in his head could do a "tomfool thing like *that*"—"that" being the forty-third consecutive example he's seen of someone attempting, for instance, flat slab construction directly on gravel pads over ice-rich ground, or some comparable idiocy. He has forgotten, perhaps, that his own knowledge was bought at the cost of perpetrating, or at least observing, a number of minor disasters himself, and that some of the principles of arctic construction he considers "obvious" took him many years to finally think through.

Fire!

Take, for instance, fire. Is there any reason why fire protection, prevention or incidence should be any different in the North where even campfires can be difficult to start in the deep cold?

Let's reflect a bit: a fire in the North, once started, may be plenty hard to fight. For one thing, water in the lavish quantities required by fire fighters is hard to come by, and pipes conveying it may even be frozen. Frostbite (a nontrivial ailment, by the way) is a common hazard among them, and larger-than-normal fighting crews are needed to haul water, to watch others for signs of freezing, and to rotate crews to warm-up areas, and so forth.

For another thing, the consequences of having a building destroyed in the North are more serious than elsewhere: a householder dispossessed by fire when there's no place else to go during the cold arctic night could be in real trouble. As this is written, the temperature is minus 50°F outside, and the late news tells of one of our major Alaskan towns losing its chief source of electricity because the power plant burnt down—the fourth such disaster in Alaska in recent years.

People dependent on electricity for keeping water liquid and running (as all people are who have running water in the North) may be forgiven a twinge or two of panic at the thought of fire, even if it is remote from their own dwelling.

Further, fire *burns* better in the North: things made of wood and organic fibers are much drier in the Arctic even than in warmer deserts. Cold air, guaranteed very dry, when warmed to room temperature has a relative humidity frequently less than 10 percent. Rugs, furnishings, and structural parts of a building therefore dry out to perhaps *half* the residual moisture of comparable things elsewhere. So they sometimes act like tinder and are easy to ignite despite the cold.

Not only that, but the really important cause of fire in northern towns is, indirectly, cold itself. For water pipes and drain lines freeze, and frozen pipes invite thawing with all manner of open-flame devices. Blowtorches, weed burners, camp stoves, even rags soaked in gasoline have thawed pipes—and in the process have destroyed dwellings. For all these reasons and doubtless more that you yourself can add, the arctic builder must take fire even more seriously than his more southerly counterparts.

Other Redundancies

No independent structure should be without "Plan B." If a furnace fails, what then? Must the piping system be drained, the traps filled with antifreeze? It's a good idea to have an alternate stove for those periods when the "old dependable" is on vacation. For small, compact structures, a fireplace helps. Otherwise, spare heaters, independent of the main ones, should be used.

The same goes for power supplies: there ought to be an auxiliary power source. The consequences of being without power when the primary system is "down" are pretty dismal. A power outage in an unprepared community not only breaks water pipes and drain taps, it destroys cherished things like house plants and tropical fish. Most people have to abandon heatless, powerless homes altogether; it's hard to convert a "modern structure" to a cozy cabin with outhouses and water buckets. Picture, for example, the northern resident whose automobile depends on an electric heater for the warmth required for starting. When the power's gone, so may be the means for self-rescue.

How About Foundations?

Heated structures over ice-rich ground had better be thermally isolated from it. For most buildings, this means elevation, so that heat losses through the floor cannot melt the ice. Some buildings (hangars, for instance, and industrial plants) carry such heavy loads that elevation on pilings is impractical. In such cases, thermal isolation by mechanical extraction of heat is required. Beneath the structural slab must be insulation, and between that and the ground will be some system to carry away the heat that penetrates through the insulation. Make no mistake about this: *insulation alone will not keep ground ice from melting;* there must be active heat extraction as well.

Here is a list, doubtless incomplete, of things the arctic builder ought to consider:

1. Go heavy on fire warning equipment: smoke detectors, periodic inspections, good housekeeping. And be sure there are lots of hand extinguishers around and that people are familiar with their use.

2. Keep a supply of firefighting water in a place were it's always ready. Water trucks, full and startable in heated garages, are an absolute must for dispersed arctic communities. If drifts or swamps or loose beach sand are a problem, the trucks might well be track-mounted.

3. Be sure each structure has ample escape routes: snowdrifts, caulked-shut doors, super-small windows—these all have trapped people in burning buildings.

4. Avoid inaccessible piping. It's the pipe that's hard to get at that invites the at-arms-reach blowtorch artist. Be sure that *all pipes* that are not easily accessible are provided with electric or other "heat tracers."

5. When pipes are frozen in spite of all, encourage *safe* ways to thaw them: not the untended heat lamp, but the slow heat of "heat tape." Not a flame, but an electric hair dryer. One of the quickest and best ways to thaw a frozen pipe is to direct room temperature air on it with a vacuum cleaner arranged to "blow."

6. Try to have duplicate facilities in structures well separated; if a structure burns, there should be another nearby which can accommodate displaced people and provide them with food, warmth and communications. Save their lives. Give them time for rebuilding or rescue.

A.
Plain box

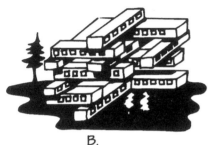

B.
"Habitat" style

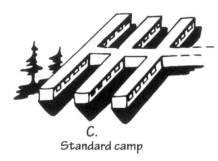

C.
Standard camp

Figure 1. From an energy conservation standpoint, the plain box (a) is by far the best, followed by the standard camp (c). The beautiful "habitat" (b) approaches irresponsibility in a cold climate. A sphere would give the best possible ratio of surface to volume, but might be impractical for other reasons.

---

1 I've made the mistake of thinking I could educate users to "do the right thing" so often that I wonder if I'll ever learn! Build 'em so that they'll survive no matter what the users choose to do. For users will come and go long after you and I have moved on.

2 Keep It Simple, Stupid.

Naturally, foundations in ground that is *not* ice-rich are no more difficult to design than those in temperate or tropic latitudes.

"Psychology" Destroys Buildings

Elevated buildings have cold floors unless the designer is particularly careful. Cold floors often prompt the disgruntled occupants into "skirting" the building by walling-in the underfloor space to keep the cold out. If the foundation has to be frozen to be stable, skirting can cause foundation failure. The cure: make sure the floors are *not* unacceptably cold.

In a related lapse of judgement, designers often specify that a space (aircraft hangars and warehouses are great for this) be kept on the verge of freezing; for any considerable heat in the structure would imperil the foundation by thawing the ice beneath. Don't be tempted: sooner or later (sooner, in my opinion) users will elevate the temperatures to what they conceive as a comfortable level, even if they have to alter and augment the designed heating system. They'll first curse the designer for not providing enough heat, then confirm their low opinion of his ability when the foundation fails. To keep an unsullied reputation, you'd do well to assume that the users will act like humans; that is, they'll do what they *think* is convenient, *not* what the foundation requires.[1]

The great hangar at Thule, Greenland, for example, had its foundation designed by astute engineers who carefully calculated how they could, by overcooling the subgrade in the winter, limit destructive summer thaw without mechanical refrigeration *provided* the hangar was kept at a fairly low temperature as well. Their plan almost worked. It would have, if people did not behave the way they inevitably will.

Do Shapes Matter?

When heat loss is a serious matter (and it *always is*), the ratio of surface-to-volume is a controlling parameter. For a given volume of heating space, the one with the smallest surface area is the one with the least heat loss. The idea, somewhat simplified, is illustrated in Figure 1.

It is easy to see that, if the building were somehow "subdivided" into modular-style units, the exterior surface area would be increased nearly four times (actually 3.67 times), and the heat loss due to conduction would increase nearly four times. It's as if your monthly heating bill were to rise from $100 to $367. Of course, the difference isn't quite that dramatic because not all heat losses are a function of area. All the same, though, there's a lot to be gained by combining spaces in such a way as to limit surface area. The "Habitat 1967" concept of tacking together individual apartments is a good way to ensure a maximum of heat-loss cost. It's worth noting that the capital cost goes up too, since outside walls are more costly than interior ones.

Beware the Fancy Solution

When it comes to most design work, the "KISS" concept[2] has some built-in genius. Example: when a water well is driven through permafrost, the water, if not kept moving, could freeze solid. Some well-meaning engineers reasoned that when water is not being extracted, it would be possible to dribble back some of it at a reduced rate. Sophisticated and careful calculations revealed that a dribbling system would indeed do the job. (The analysis took into account the frequency of water use, the required extra volume of storage needed, the maximum duration for water not in use, and all.) How would you feel about this? If you were the designer, you'd feel proud of the simplicity and economy of your solution to a tough problem—until you discovered that some bright soul "fixed the leak," and the well became solidly frozen, all the way down, and there was no provision for a fast and easy thaw.[3]

And Now, Let Us Explore

The assembly of "small ideas" you've just read are ideas left out of the articles that follow. Some of them simply don't justify a whole chapter by themselves, and others had to be mentioned *someplace*, but there was no appropriate place elsewhere.

A few of these ideas you will find repeated, perhaps too often, in the pages that follow. Forgive me for that.

My aim is to provide a basis for the exercise of common sense in design for the North—and "common sense," as we all know, becomes "common" and "sensible" only to those who take the trouble to understand the basics: what makes some things work and why, when all else seems to fail.

A designer with common sense bought it by experience. Experience is what you get when you profit from mistakes. Mistakes are what you make when you haven't enough common sense; maybe these chapters will alert newcomers to experiences others have had, that we may someday enjoy together that uncommon virtue, "common sense."

—Eb Rice

---

3 This is a true story. It cost much more to thaw the well than a simple heat tape would have cost in the first place. And one had to be installed in the end, anyway.

Permafrost: Its Care and Feeding

CHAPTER 1

Introduction

Permafrost is one of those casually coined terms designed to be used until the "right" word comes along, but whose popularity has become so great that is fixed in both the technical and the popular vocabulary of the world. *Permafrost* has turned into the "right word" despite repeated efforts to replace it with a better one.

I suppose the fruitless efforts to abandon the term are brought about because the *perma* implies foreverness, which is not a necessary condition, and the *frost* implies ice crystals and cold air, which gives a sense of chill, maybe, but fails to give an impression of the ground the term is intended to describe. My own unabridged dictionary doesn't show the word "permafrost" at all, but a later College edition has a wholly unsatisfactory definition: "subsoil in the Frigid Zones that remains permanently frozen to depths of over 1000 feet." For now, we will define permafrost as "perennially frozen ground" and go from there.

Permafrost: What is It?

Perhaps the most widely recognized definition today is ground that has been below 0°C for two winters and the intervening summer. If the ground is dry sand, it is permafrost no less than if it were a conglomerate of soil particles cemented by ice. From an engineering standpoint, frozen soil *sans* ice is nearly as tractable as similar soil in temperate or tropic regions. Unfortunately, much frozen ground tends to be cemented by ice; it also may be interlarded with large ice masses whose melting could bring about subsidence, erosion, and structural distress.

Where is It?

In general, people associate permafrost with the polar regions, and it lies as a thick layer (under a thin "active layer" of seasonally melting soil) in the very high latitudes, thinning gradually toward the lower latitudes, until the layer becomes discontinuous, then sporadic, at its southernmost boundary. On mountain peaks, it is possible to find permafrost near the equator, but permafrost south of the "bush country" of Alaska, Canada, and Siberia is rare. Even Scandinavia, despite its high latitude (up to 72°N), has permafrost only in its high, isolated areas and in some sheltered, wet, boggy ground. Yet the continuous layer of frozen subsoil extends far southward in the "continental" parts of Canada and Siberia, all the way down to 53°N. Trees can and do grow over frozen ground, but the "tree line" marks fairly well the southern boundary of the *continuous* layer of permafrost. It is not far wrong to say that the treeless tundra overlies continuous permafrost, and the forested taiga covers the frayed margins of the permafrost blanket. In the interior of Alaska, in the taiga of the Yukon, Kuskokwim, and Copper

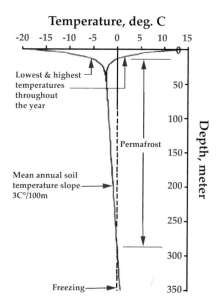

Temperature, deg. C

Lowest & highest temperatures throughout the year

Permafrost

Mean annual soil temperature slope 3C°/100m

Freezing

Depth, meter

Figure 2a. Average and bounding temperatures at various depths.

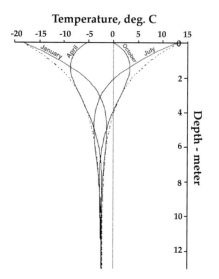

Temperature, deg. C

January   April   October   July

Depth - meter

Figure 2b. Seasonal temperature variations as a function of depth.

River valleys, the condition of the permafrost is marginal. Almost at the melting point, permafrost exists or not in these areas depending on such apparently minor influences as whether the land slopes to the north or to the south, whether the forest cover has remained intact for many years, or whether the land is swampy. Permafrost occurs sporadically even around Anchorage and on the Kenai peninsula.

Why is Permafrost?

If the climate is cold enough, for long enough, there will be permafrost. You could even create your own by keeping the ground surface cold for a long time; the frozen volume you could build up would be indistinguishable from naturally occurring permafrost.

Deep in the earth, heat is generated, probably through nuclear decay. This heat flows to the earth's surface to be lost eventually in outer space. The surface temperature of the world is determined by a complex set of interactions between solar radiation, radiation to and from the sky, and the near-surface conditions established by clouds, wind, rain, water, and terrain.

The heat flow from the interior to the surface of the earth results in a remarkably constant temperature gradient throughout the earth—about the same in the Arctic as in the tropic regions. Apart from volcanic areas, the gradient is somewhere about 3 Celsius degrees per 100 meters. If the soil surface temperature averages, say, 10°C, at a depth of 100 meters it could be expected to be somewhere around 13°C, and so on. Wherever the average soil surface temperature is a degree or so colder than freezing water, permafrost will be present in the soil, and the colder the average surface temperature, the thicker the permafrost layer will be (Figure 2a). Of course, in reality, this conceptual picture is muddied by differences in materials, long term changes in climate, proximity to lakes and streams, elevation, exposure to the sun, and so on.

Several local factors and seasonal temperature variations complicate the temperature in the first 10 to 20 meters of the surface soil throughout the year. For instance, if the surface temperature suddenly became elevated, the temperature curve would bend sharply toward the right near its top, as the surface warmed and thawed. Since it takes time and heat to warm up large masses of soil, the soil at any depth is delayed in "getting the word." This happens every spring. By the time the thaw penetrates a short distance into the frozen ground, it may be autumn again, and the surface could be freezing once more. As winter continues, the curve bends sharply to the cold side as the surface temperature goes lower and lower, while a few meters down the soil is still warming with the previous summer's heat. Figure 2b shows the effect of seasons on the temperature curves. A moving picture of the curves would show the surface temperature fluctuating wildly between late-summer temperatures and cold late spring temperatures. A few centimeters below, however, the temperature is never quite so hot, nor quite so cold, as the surface; and greater depths show even smaller fluctuations, until at a depth of 10 or 15 meters the fluctuations are too small to measure conveniently. These instantaneous temperature curves are called "whiplash curves" because of the way they lash to and fro as the seasons change. The upper and lower limits of temperature, warm and cold, for the soil are asymptotic to the mean temperature gradient, and can be called the "trumpet curves."

From such a simple plot as Figure 2b, one can derive some surprisingly important insights. For example, it is apparent that for the site shown (which is in Fairbanks, by the way) the soil is thawed at some time during the year at all levels shallower than about 2 meters. The top 2 meters, therefore, are the "active layer"—sometimes frozen, sometimes not. Below the active layer is the ground that never gets warmer than 0°C, permafrost. At this site, you can see the permafrost is also *not much colder* than 0°C; it is almost thawed. Any little thing raising the surface temperature a degree or so would, in time, melt the permafrost completely.

It is also apparent that, on average, the active layer *must freeze clear down to the permafrost every year*. If it should fail to do so for several years running, the permafrost will begin to melt from both the top and the bottom. If this situation continues, the permafrost layer will continue to grow thinner until a new equilibrium forms, i.e., if the soil surface temperature were to remain warmer than before. This fluctuation could be caused by a change of climate, a clearing of vegetation, or by cultivation. A forest fire, for instance, by removing the vegetation cover, can alter the thermal regime for years—until the original growth becomes re-established. What, then, of the supposition that permafrost is a fossil relic of the ice ages? Some of it could well have been frozen since the Pleistocene, but it is unnecessary to postulate ice ages to explain permafrost, and to do so hides the truth. Barring a few anomalous cases, permafrost is a product of equilibrium with the *present* climate.

In places like Fairbanks, where permafrost is marginal and near thawing, a location on a north-facing slope will overlie a deep layer of permafrost, whereas permafrost may be entirely absent under a slope facing south. Swamps tend to be frozen very deeply, with only a few inches of thaw during the summers. Well-drained level ground around Fairbanks may or may not have permafrost, depending on such subtleties as the vegetation of its recent past. Mere clearing, in some instances, will change the surface temperatures enough that the icy ground will begin to melt. If it is highly ice-filled, many years will be required to reach a new equilibrium.

From Figure 3 we can also see that the thermal situation in Barrow is not at all as precariously balanced as it is in Fairbanks; only a little of the surface thaws during the short summer. It takes a very great change indeed to cause serious melting below the top meter of soil. The frozen ground is in little danger of thawing unless someone places a heated slab on it, in which case there can (and have been) spectacular deformations. Anchorage does not have permafrost, but only seasonally frozen ground.

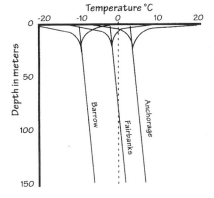

Figure 3. Typical ranges of temperatures in Alaska.

Whence Cometh All that Clear Ice in the Ground?

When ground of silty texture freezes, frequently moisture migrates to the freezing surface. While no one perfectly understands this mechanism yet, to oversimplify the phenomenon, one can say that by vapor transfer migration through voids and liquid migration along grain boundaries, masses of ice can begin to form along the freezing front. This process causes a drying effect in the adjacent soil. If water is available, and if the soil has capillarity (that is, if it can act like a wick), the moisture can migrate to the freezing front. Considerably thick ice layers (1 to 100 mm are common) may develop, displacing the soil upward to make room for the ice. This is the familiar "frost heave" that bedevils engineers and others wherever soils are fine grained, damp, and exposed to freezing. Frost heave is the prime cause of displaced house porches and steps, highway pavements, and retaining walls.

What About Ice Wedges?

Ice-rich permafrost (the only troublesome kind) got some excess ice by this lensing process when it originally froze. But there's another mechanism responsible for the truly massive volumes of ice in perennially frozen ground. Everyone who has flown over the Arctic in the summer has been impressed by the vast areas of ground that appear to be divided into a more or less regular pattern of rectangles or hexagons (See photo). Sometimes the polygons have raised edges and ponds within. Sometimes the centers are high and the edges low. Always the boundaries prove to overlie masses of ice, wedge-shaped in cross-section, with the point of the wedge down. If the ice could be removed without disturbing the surrounding soil, the result would be a network of interconnected wedge-shaped tunnels, all with their flat tops located at the depth of the annual thaw. These tunnels could be as much as 3 meters wide at the top and perhaps 10 meters deep (Figure 4).

Now it seems obvious to us how these polygons were formed with their ice-wedge boundaries. The ground contracted when frozen hard

Ice-wedge polygons, ad infinitum.

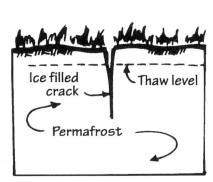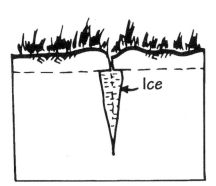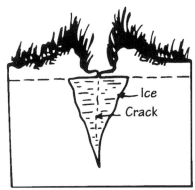

Figure 4. Ice wedges start from small cracks. In time, ice wedges cause a network of polygons in the ground. During summer, all ice above the thaw level melts. Each winter, the ice wedge cracks open in the center.

and cold in the winter, and the cracks filled with snow and frozen meltwater. The ground re-expanded as it warmed (but did not thaw) during the summer. It could not expand into its former cracks, since these were filled with ice. So it expanded (forming either the raised edges or humped backs in the polygons). As long as the ground never thawed, the ice-filled cracks endured, and the following year, new cracks formed. Interestingly, the new cracks usually occur right down the center of the ice in the old cracks, and the wedges widen each year. Ice wedges can grow to startling dimensions. Since this process could conceivably go on forever, the dimensions of active ice wedges are an important clue to their age, and possibly also to the age of the current cold phase of the climate.

The mechanism of ice wedge formation was not always so plain. Such acute and patient observers as Nansen and Stefansson attempted unsuccessfully to explain the massive ice exposures they observed in eroded shore lines. Stefansson, in fact, postulated that sea ice, under pressure created by wind and current, pushed its way like a bulldozer under the tundra edge. An ingenious thought, perhaps, but wrong.

There are a few places left in the world where ice wedges have grown so wide that their upper parts have coalesced, and there is clean ice everywhere under the tundra (Figure 5). Except for the explanation just given (originally by Leffingwell, in 1910) scientists would still be

Figure 5. Sometimes ice wedges seem unable to stop growing (Novosibirsk). Nansen observed continuous solid ice where apparently ice wedges have grown large.

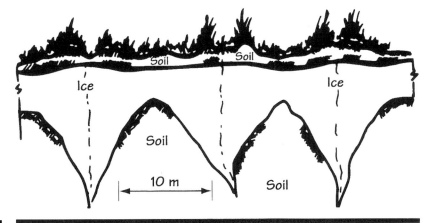

hard-pressed to explain how all that dirt and vegetation managed to form atop a continuous layer of ice.

The polygons vary in size according to the tensile strength of the ground, but generally range from about 8 to 18 meters across. Incidentally, the same kinds of polygons occur in temperate zones as well. But

Near Fairbanks, it is common for even perfectly constructed houses to encounter the embarrassment of having been built over ice wedges deep underground. This one has a differential settlement greater than one foot. (The trees remain vertical!)

where there is no permafrost, the cracks are reclosed each summer. Since they are usually hidden under snow in winter, temperate zone polygons are rarely noticed except for their surface manifestation as they cross roadway embankments. (These we carefully preserve with asphalt, including even the upset edges, doubtless to remind us of the way frost cracking plays for keeps in permafrost areas.)

Many people do not realize that the Alaskan interior has vast areas of ice wedge polygons under the spruce forests. These show up in a matter of 15 to 20 years after the forest is cleared, leaving fields filled with humps so high that farm machinery cannot operate until releveling has taken place. Such ice wedges in marginal permafrost often go unsuspected, since they are masked by the very deep (circa 5 meters) active layer and heavy vegetation cover.

Many heated structures have been innocently built on such ground (see photo). After a few years, deformation may become too great for even the most flexible building structure or homeowner to tolerate. A conservative estimate of the cost to date for the premature destruction of houses and other structures on ice-rich ground is millions of dollars in the Fairbanks area alone. As many as 300 homes in the Fairbanks area are currently undergoing foundation distress due to melting permafrost. And yet, with competent engineering advice, such ground may be used acceptably for structures of any size or complexity.

Of Ponds and Pingos

Ponds and lakes deeper than a couple of meters do not freeze to the bottom, so geothermal heat is interrupted in its path to the surface,

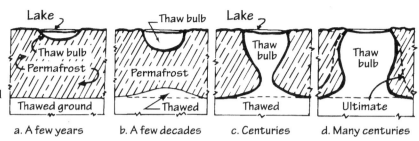

Figure 6. Stages in thawing under a newly formed lake in a permafrost region. If the lake (or reservoir, heated building, etc.) is large enough, the thaw may extend all the way through the permafrost.

a. A few years   b. A few decades   c. Centuries   d. Many centuries

and each such pond comes equipped with an unfrozen "thaw bulb" (Figure 6). The size of such a bulb depends on the rate of heat flow and the climate (the thaw bulb is smaller where the climate is colder). In the North American Arctic the depth of thaw under medium-sized lakes can approximate the width of the lake.

Large lakes (over 400 meters wide) are likely to have a thaw bulb that completely penetrates the permafrost—see Figure 6c—so that under such a lake there should be no massive ice in the ground, and consequently no disastrous settlement due to melting. (Of course, one might have disastrous settlement due to soft mud, but the care and cure of that problem is hardly different in the Arctic than in Louisiana.)

Some lakes are newcomers. These may not have been in place long enough to establish their eventual thermal balance, and the bottoms may not yet have had enough time to "thaw through." Conversely, when a lake is silted full, or freshly drained, its thaw bulb remains, steadily diminishing in volume, to testify that once upon a time there was a lake. If such a thaw bulb is surrounded by impervious material (read "permafrost") the freezing water will find its expansion room limited as it turns to ice. A slow fountain may result, freezing as it ponderously raises a blister on the earth (see Figure 7). Such blisters—ice mounds with a topping of soil—are called "pingos," and there are thousands of them on the North Slope between Prudhoe Bay and Amundsen Gulf, with the most spectacular specimens near Tuktoyaktuk, east of the Mackenzie River delta. In flat areas, pingos (which can be up to 50 meters high and

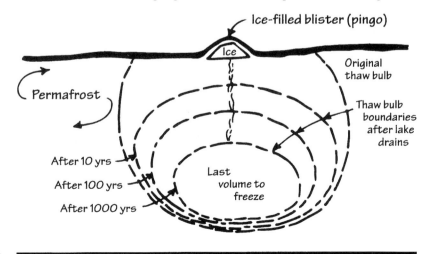

Figure 7. Ice occupies more volume than it did as water. When water is frozen in a closed container, therefore, "something's gotta give." In some areas, pingos form.

This pingo, near Tuktoyaktuk, NWT, is one of the largest in the world.

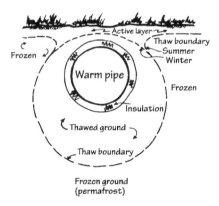

Figure 8. Buried warm conduit.

1 Oddly enough, small pingos, called "open system pingos," form in Alaska's interior valleys by an entirely different mechanism. There are several of these in the Goldstream Valley, within a few miles of Fairbanks. For a brief, interesting discussion, read A. L. Washburn, "Periglacial Problems," in *Field and Theory: Lectures in Geocryology*, edited by Church and Slaymaker (1979).

This photo shows three ice wedges exposed in a cut. It is a rather spectacular example of this naturally occurring permafrost feature of the Arctic. Photo by Mikhail Kanevskiy.

150 meters across) provide the only high spots in the landscape, and have long been used (although ill advisedly) as sites for geodetic triangulation stations (see photo[1]).

Other Anomalies

There are other kinds of ice formations, such as buried lakes, ponds, and streams, not to mention buried chunks of glacial ice. The photo on the left shows one example, a chunk of ice embedded deep in the soil. A good Arctic engineer must expect to deal with all these formations in one way or another.

What Do You Need to Bury a Hot Pipeline in Permafrost? Plenty of Luck!

Occasionally builders bury warm conduits—utilidors, water lines, sewers, oil wells, and so on—in the frozen ground.

Around such a conduit there will be a thaw bulb (more accurately, a thaw cylinder) regardless of the amount or kind of insulation supplied (Figure 8). This inevitably makes things interesting for engineers. If the ice-rich ground is newly thawed, the resulting mud could flow down the trench, leaving the pipe uncovered and exposing a gorge to further erosion. Thoughts like this provide topics for endless imaginative speculation. In some communities the difficulties of maintaining a warm tunnel in frozen ground have impelled engineers to decide on above-ground utilities. The cost of doing so is also high, and the subtle social costs and inconveniences may even outweigh the physical difficulties of the buried system, as we will discuss in later chapters.

One special consideration in the pipeline was the oil well—a vertical tube, filled with oil moving fresh from the hot depths. Such a tube must develop a vertical cylinder of thawed ground around it. Trouble from this setup can develop in two ways: a producing well may melt a cylinder of earth containing massive ice whose conversion to water leaves the overlying appurtenances without support. Or if the well is inactive,

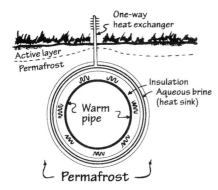

Figure 9. Buried hot pipe with heat sink.

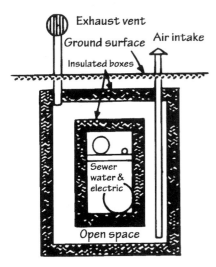

Figure 10. Double-walled utilidor design.

"freeze-back" may take place, squeezing the line with the same kind of hydrostatic pressure that in a different context produces pingos.

There are many ways to combat these difficulties—about as many as there are engineers to think of them. Actually, the problem can be solved by an astute combination of insulation and "heat sinks," but the solution is both intricate and expensive. There have been a couple of extremely interesting solutions concerning the Trans-Alaska Pipeline (Figure 9). The trick is to extract enough heat during the cold months to freeze a heat sink composed of brine whose melting point is a few degrees colder than the thaw temperature of permafrost. During summer, this heat sink undergoes melting, and carefully designed systems should have enough capacity to weather a full summer before the entire heat sink melts. By fall, the heat sink should be nearly melted and ready to "recharge" throughout the winter. Perhaps a more common technique is to provide a double-walled insulated box in which cold air is circulated between the walls (Figure 10).

So What's the Rub?

There's no rub. If permafrost can be kept frozen, it can be treated like any other rock, with due regard to its physical character. If it cannot be kept frozen, well, that is a problem.

In the high Arctic, the permafrost is cold and insensitive to casual disturbance, hence it makes an admirable foundation. Where permafrost is marginal, more careful design is indicated. Instead of the credo, "Keep it frozen," which works so well in the high Arctic, one must select between: "keeping it frozen even if you have to refrigerate," or "keeping it thawed by melting it out to begin with." Watch out for an aggressive and intractable active layer.

In later chapters, we will explore foundations designed especially for heaving ground, the problems of handling sanitary water supply and waste disposal systems in the Arctic, and other special problems of polar communities.

References

Burdick, John L. (Ed.). *Symposium on Cold Regions Engineering* Vols. I and II. Dept. of Civil Engineering, University of Alaska and American Society of Civil Engineers, 1971. (Contains many articles of significance to engineers dealing with permafrost).

Church, Michael, and Olav Slaymaker. *Field and Theory: Lectures in Geology.* University of British Columbia Press, Vancouver, 1985. (Contains several interesting chapters on geological ice formations.)

Lachenbruch, Arthur H. *Some Estimates of the Thermal Effects of a Heated Pipeline in Permafrost.* USGS Circular 632, Washington, D.C. 1970.

Leffingwell, Ernest dek, *The Canning River Region, Northern Alaska.* United States Geological Services, Professional paper 109, Wn. 1919.

Nansen, Fridtjof. *Farthest North, The Voyage and Exploration of the Fram, 1893–96.* Vol. 1. Constable, 1897.

Stefansson, Vihjalmur. *My Life with the Eskimo.* Macmillan, 1924.

Northern Construction: Siting and Foundations

CHAPTER 2

Introduction

Many years ago the inhabitants of the North included mostly Eskimos, Indians, Lapps, and various Siberians. Their dwellings were admirably adapted to their ways of life, and some of these structures exhibited ingenuity and a command of basic scientific fundamentals that deserve study and emulation today. Now, however, as "southern people" move into the north lands, and as northern people seek to avail themselves of the culture of the 70°F environment, dwelling designs must incorporate special parameters.

Stefansson, during the 1920s, bemoaned the fact that northern American beaches, once abundantly supplied with useful driftwood, had become barren. Old-style houses, semiburied and warmable by the heat of the occupants and their cooking lamps, were being abandoned for fuel-hungry frame dwellings like those of the newly come traders, missionaries, and government men. To people experiencing a transition from nomadic life to settled existence near church, school, and store, mining the beaches for fuel was but a small price to pay for a house that could be lived in all year round—one whose floor would not turn to mud with the coming of spring, and one that could have real windows. No matter that firewood was becoming scarce.

No matter that the new houses cooled swiftly and bitterly when the fires died. No matter that drafts were frequent and biting: this was the way of the affluent newcomer, and what he could endure, all could endure for the sake of status, style, and year-round occupancy. So people, Native and newcomer alike, contrived to make their homes in uncomfortable, costly, and inadequate dwellings.

Neither newcomer nor Native seemed to be aware that the temperate zone dwelling was not always adapted to the Arctic. Indeed as a fervent pioneer once said, in an attempt to persuade me of the futility of striving toward comfort and economy, "When you hire out to Alaska, you hire out to be tough." It was unworthy, he figured, for a dweller in the Last Frontier to be thinking of comfort. And as for economy, one would be amazed at how much fuel he could save if he never removed his furs and mukluks. Stefansson probably was giving much the same message: if one lived in the old ways, toughly, there would be plenty of driftwood for all.

But frontier toughness is in for a change. The new northern resident usually demands and gets the comforts normal to the temperate zone: flush toilets, heated baths, centrally heated houses, hot running water, and even an attached garage. This is thought to be merely our due. If people live like this "outside," the feeling is, we should live no less "well" in the North.

Trouble is, people developed these comforts and conveniences for temperate places over many decades. To reproduce them in the northland without knowledgeable modification sometimes results in total failure, and always in extra, unnecessarily high costs, now and forever. I say "unnecessarily" because careful attention to the particulars of this environment can minimize the inevitably high costs of building modern dwellings in the North. And northerners are properly becoming impatient with the inadequacy and costliness of the hasty and carelessly adapted temperate zone house.

Siting: Sleep, Snowdrifts, and Sun Worshippers

Choose the Site Wisely

A builder must take special care in locating the arctic dwelling for many reasons. For example, in warmer regions the sun rises high overhead on every cloudless day; in the polar regions the sun is never high, and in winter it may not even rise at all. In the Alaskan summer, it rises in the north, moves all the way around the house, then sets in the north—if indeed it sets at all.

A site that fails to make maximum use of the existing sun angle is therefore one to be shunned. This is true in the subarctic as well. Even there, the difference between south-facing slopes and other terrain is pronounced. The variety and health of vegetation on sun-facing slopes attest to the favorable climate to be found there. In the Fairbanks area— typical of the "zone of discontinuous permafrost"—slopes facing south are usually without permafrost. Other slopes, together with poorly drained flat terrain, are often underlain by ice-rich ground. Such sites are possible building sites, to be sure. But for reasons of psychology (one can see the sun in the south) as well as terrain, the south slope is preferable and cheaper, too, in the long run.

Of course, a site selected in the North, as elsewhere, should be conveniently close to water, to transportation, and to a place of work.

To the true Alaskan, the dog team or snowmobile can go anywhere with enough effort, but to the rest of us it is best to scout out which roads are plowed, by whom, and how often, and how treacherous the road may be after plowing. Even with moderate maintenance, some intersections are best left to dog teams throughout much of the winter. Then again, some roads are easier and more comfortable to traverse as ice-covered skating rinks in the winter than they are as mud baths or dusty washboards in the summer. To each his own—just know what you want, and what you're willing to endure to get it. One neighborhood in Fairbanks creates a new entrance every winter, simply by driving across the frozen Chena River. The trick is to know when to stop using it each spring. Once the access constraints are satisfied, a dwelling should be placed on ground that will not become a quagmire when thawed or a pond or river during breakup or after heavy rains. It is never easy, particularly in the winter, to tell what will be flooded during breakup since this is controlled not only by gradient near the site but by the gradient and the thawing conditions a long way away.

Figure 1. Beware the "cozy" site.

Wind in the treeless tundra is a force to be reckoned with, and shelter from it much to be desired. But constant Arctic winds can deposit an enormous amount of snow, quite capable of filling a completely sheltered cove, even in areas where the total snowfall is minimal. Moral: either move to the wooded taiga where drifting is minor, or locate on exposed plains or ridge tops, then plan on snow drifting on the leeward side of all obstructions (Figure 1). Mountainous terrain may have avalanche areas, which are strictly to be avoided at any cost.

A "view," composed of mountains, lakes, rivers, or just one's neighbors across the way, may also be desirable. The cost for such a view may include wind exposure, colder temperatures, or proximity to the feeding grounds of the Alaska State Bird, the mosquito.

The relationship between elevation and temperature can be interesting in the North. We tend to associate northern latitudes and higher elevations with cooler temperatures but some areas differ. In Fairbanks, where there is very little wind in the winter, extreme temperature inversions can occur. Relative low areas can be 10 to 15°C colder than the adjacent higher areas. In some instances, this phenomenon is so extreme that your booted feet can get frost bit while your uncovered ears are still warm. Small river valleys and slough channels tend to be very cold, while high spots on south-facing hillsides are much warmer.

Which Way Should a Dwelling Face?

Woodsmen traditionally placed their tents so that the door faced east—the rising sun was expected not only to wake the occupants, but to erase the night's chill. Religious structures were also oriented with their alters to the east (and thus their main entrances to the west). Mosques are supposed to face the Kaaba in Mecca. Most dwellings, if given ample land, are placed facing the "view," or at least facing downhill. What about the arctic house? Does it matter which way it faces?

This topic should perhaps remain a matter of personal preference, as well as a matter of semantics about what constitutes the "face" of a building. There are, however, some pertinent parameters.

The first sun in the arctic spring, for instance, appears briefly at the southern horizon. Alaskans watch with amazing eagerness the daily lengthening of the sun's path. No one who has lived through the arctic winter will discount having windows on the south, with perhaps a sheltered outdoor area as well, to accommodate springtime sun worshippers. Conversely, newcomers tend to want few openings on the north side of a dwelling, for they prefer to sleep undisturbed by the horizontal rays of the summer's midnight sun. Most Alaskans prefer to place the side with the best windows toward the view (south), downslope (south) and capable of admitting the noon sun (south). But we like at least a few windows facing all the other directions too.

We northerners tend to dwell on the winter cold, probably for the same reason your grandfather used to tell of walking 20 miles through the blizzard to get to school every day. However, the Alaska interior can be equally hot (25° to 35° C) in the summer; hence some shade and a way to close off the south windows is also desirable. Another engineer always recommends one more design tip: Place a window facing any direction from which a bear might come, and be sure to include a door on the opposite side.

Shovel, Shovel, Toil, and Trouble

Snow drifting can be important, for snow drifts are not merely marvelous sculptural elements: drifts can become frustrating or even disastrous at an arctic site. In the high Arctic, winds blow continuously over the treeless plains and winter lasts for many months. Snowdrifts are not limited, therefore, by the availability of snow or by the wind of a single storm. Snow can accumulate in virtually unlimited quantities, and if one must dig the roadway or outbuilding out of a drift early in the autumn, the project may continue every day, all day, until late spring. It is vital to locate roads, entryways, and storage areas away from the downwind side of any building that obstructs the wind's free flow. And the building should, if possible, be aligned with its narrow-

Figure 2. Snowdrifts make a difference. Site plan (b) is laid out to minimize drifting, and maximize access. Prevailing storm winds are from the left.

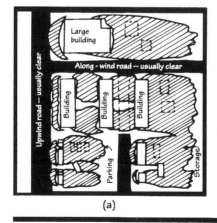

(a)

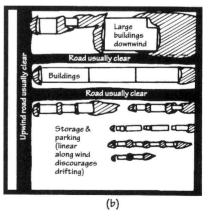

(b)

est face toward the prevailing wind. Even vehicle parking must be carefully arranged to discourage drifting (Figure 2).

Are You *Still* Planning to Build in the Arctic?[1]

The Big Picture

There are two kinds of permafrost, bad and really bad, otherwise known as "thaw stable" (bad) and "non-thaw-stable" (really bad).

Thaw-stable permafrost is dry enough that if it thaws it will not settle enough to damage your structure. On the good side: It is very strong when frozen. On the bad: Every thing else. Thaw-stable permafrost will freeze your water and sewer lines. Generally impermeable, it is a poor place for septic systems, and it may affect other sorts of drainage as well. On top of that, it can be very difficult to excavate. Other than these factors, thaw-stable permafrost can be built upon just as one would build in Hawaii, except for the palm trees. Non-thaw-stable permafrost is, according to some, evil incarnate. According to others, a builder has three basic choices: avoid it, remove it, or live with it. The first and best advice is to avoid the problem if at all practical. Move someplace where there isn't non-thaw-stable permafrost. Many choose Hawaii.

But just choosing higher ground, south-facing slopes, gravel-filled river valleys, or rock outcrops may be sufficient. If, due to tenacity or circumstance, you are still determined to build on ice-rich permafrost, you can avoid the problem by digging out the ice-rich ground and replacing it with good fill or by installing a pile foundation through the ice-rich material into the material below. Builders frequently use the excavation and replacement option to eliminate permafrost to a depth of 7 meters (about 8 yards). The pile option is frequently used to 30 meters (33 yards) or more.

If you refuse to move and the permafrost is too deep to remove or penetrate, then you still have two options left. Living with it is the next easiest solution, if you can accept the restraints. There are four primary choices:

1. Design a dwelling so strong it won't bend, even if the soils heave or settle under the foundation.
2. Design the dwelling so that it can withstand some distortion and then be releveled easily when necessary.
3. Install piles deep enough so that the permafrost won't thaw to the bottom of the piles during the life of the structure.
4. Insulate and refrigerate below the structure's footprint to preserve the permafrost.

These strategies can be used singularly or together, and anything a builder can do to stabilize the permafrost in conjunction with these techniques will be advantageous. The last option is to lick the problem either by thawing the permafrost before building or by arranging to keep it frozen forever (or, more practically, for the life of the building). Thawing is generally an expensive, muddy mess. And since it is necessary to restabilize this transformed area, the remaining mud must be preloaded (compacted) before to construction. Despite the expense and

1 Or subarctic: Arctic is where permafrost is continuous (see previous chapter). In the subarctic you have a good chance of finding a permafrost-free site, in which case your problems are minor.

the mess, this technique has been used for generations, and it will work to depths to perhaps 15 meters or so. Several projects have recently used this technique in Fairbanks, and it is interesting to note that you can never get an experienced contractor to bid a second one of these jobs.

Keeping the ground frozen can be as simple as raising the house and surrounding it with northern landscaping, or as complicated as installing a ground refrigeration system. Merely raising the dwelling works if the climate is cold enough, but warmer climates or special construction needs may require artificial refrigeration. For more details on these and other foundation schemes, read on.

The Exposed Bottom

Corbusier, the twentieth century French architect, suggested that large buildings be elevated on massive columns (*pilotis*) so that the bottoms could be seen by passers-by. This, he claimed, gives a better feeling for their volume, so that the structures can properly fulfill his dictum that a house is "a machine for living." The prehistoric Swiss lake-dwellers elevated their structures on pilings above the waters, partly for defense, partly for convenience. Tropical peoples in many areas elevated their houses on posts, too, mostly for defense against rising flood-waters, but possibly also to quell the need to share the dwellings with livestock, insects, or other casual fauna. Elevated structures are also easier to cool, but this virtue has little appeal in the Arctic.

However useful or aesthetic it may be to elevate structures in other climates, in the Arctic, where the ground is so often fat with ice, it can be vitally necessary. To place a heated structure against frozen ground is to thaw it—and when ice-rich ground thaws, no structure built on it is likely to survive (Figure 3).

The essence of structural survival for anything built on permafrost is to prevent the building's heat from reaching the ground and to ensure the foundations are permanently immobilized in frozen soil. This means separating the heat source (the house) from the ground. I want to take this opportunity to dispel the idea that insulation alone will

Figure 3. The elevated structure (or, drifting and thawing can be exciting).

Somebody made a mistake.

suffice: it will not. Insulation, whatever its virtues in other contexts, can never *stop* thaw; it can only *delay* the inevitable. To repeat: *total* separation of ground from heat source is vital. Insulation alone cannot substitute. Much as we may abhor allowing winter to pass freely beneath our floors, there are no inexpensive alternatives, nor are there likely to be. While any good arctic engineer can devise a way to extract heat without elevating the surface, generally these solutions are not economically competitive with traditional small building foundations.

Why do we expose the bottom? First, in some environments, a structure elevated enough to allow the unimpeded intrusion of winter between itself and the ground will remain stable, and its foundation can be easily contrived to remain frozen in place. Second, an elevated structure requires an absolute minimum of site disturbance: no alteration of drainage patterns, no massive excavation, and no destruction of organic ground cover need occur (although some recommend a gravel pad underlain by a geotextile over the area, which will help control fire and mud). Third, the elevated structure is the only enclosure, other than a buried one, that will not promote severe snow drifting in the vicinity.

The elevated structure is, or can be, flood proof, cheap, and easy. Its simplicities commend it to inexperienced builders working under difficult conditions. If, somehow, it should fail, it can be restored simply and easily to rights. And suitable foundations are comparatively easy to transport—even by air—to most Arctic locations.

Unluckily, an exposed bottom can be cold. Maladroit designs have resulted in floors so uncomfortably cool as to encourage occupants to "skirt" structures with some kind of sheathing to protect the "crawl space" beneath the building from winter's cold. Such skirting can be used, carefully, for structures in the high Arctic where the ground temperature is quite low. But skirting can be disastrous in areas of marginally warm permafrost, such as at Bethel, Fort Yukon, or Glennallen. In these areas, the frozen ground is only a degree or so colder than the thaw point, and there is very little reserve to accommodate heat added to the normal summer flow. Again, a thaw bulb will form, which ultimately may swallow the foundation. Dramatic failures have occurred to foundations that we initially well designed and constructed, but which the owners skirted to improve comfort, utility, and their sense of esthetics.[2] If the floor is well insulated, the skirting can be beneficial during summer, protecting the space from heat and helping keep the temperature under the structure cool. But the chore of removing and storing the skirting every winter is too much for even the most diligent of homeowners, and before long it remains in place all winter with disastrous results.

So How Do We Do It?

Pile foundations and post and pad have been used for centuries in all parts of the world, and both are often used in the Arctic. Modern engineers have developed more sophisticated methods that may be more economical or reliable under some circumstances.

One is a three-dimensional matrix of steel tubing called a Triodetic foundation, which we will discuss in more detail later. You may have seen something similar supporting the roof of some large buildings.

The second is an inverted adaptation of the anchors used to guy power poles all over America.[3] The anchors are screwed into the permafrost and the foundation is placed on clips attached to the drill stem. They can be installed by hand-operated equipment, and the foundation can be releveled easily using jacks made to fit between the drill stem and the foundation clip (Figure 4).

Shake, Rattle, and Roll

Alaska is one of the most seismically active areas in the world. There are literally thousands of earthquakes every year, and out of those, Alaskans report feeling something around 150 a year.[4] Unless you simply enjoy the thrill of the ride and don't mind your china racing around the shelves, your home should be stabilized against earthquake movement as well as wind force. This is particularly important with elevated foundations, since many are not designed to bear horizontal movement.

The Bottom Line

To summarize: despite the cold floor hazard, many virtues recommend construction on a raised foundation for arctic building. Among these, two are especially salient:
1. the heat segregation inherent to elevated construction prevents leaking heat from the structure into the ice-rich ground, and
2. elevated foundations allow arctic winds to flow unimpeded, discouraging serious snow drifting.

Together with simplicity, flood-proofness, economy, and environmental suitability, these make thermally isolating a dwelling by elevation a nearly unbeatable technique for construction over ice-rich terrain.

The Ultimate Ground Cover

Tundra is the perfect arctic ground cover. It protects permafrost despite the climate's temperature extremes. Tundra is high enough to create shade in the summer. It traps evaporating water, sucking it up and spitting it out to the atmosphere (evapotransporation). For every pound of

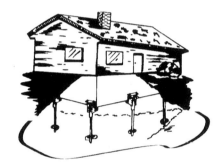

Figure 4. A helical anchor foundation.

2 I regard this as an architectural or engineering failure: if the design leaves the occupants so uncomfortable that they are impelled to actually do something, the failure is no less serious than if the foundation design had been structurally and inherently faulty.

3 Alaska Foundation Technology, Sitka Alaska.

4 University of Alaska Fairbanks Geophysical Institute Earthquake Information.

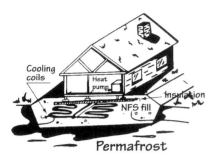

Figure 5a. Mechanical refrigeration.

Figure 5b. Air ducts used to stabilize permafrost.

Figure 5c. Single-phase thermal syphons.

Figure 5d. Thermal syphons beneath a house foundation.

water expelled this way, 1000 BTU's of heat energy go along with it. This tundra-trapped water acts as a one-way heat valve, allowing heat to leave the ground more easily in the winter than it enters in summer.

Every time we disturb the natural vegetation in the Arctic we cause a net increase in annual ground surface temperature. Throughout much of interior Alaska, the average ground surface temperature is a few degrees warmer than the average annual air temperature, a few degrees below freezing. If we disturb the ground and thus raise the temperature even a little, the temperature goes above freezing and the permafrost will thaw.

Landscaping in the Arctic is like being on a diet: nearly everything natural is good for the dwelling and nearly everything unnatural is bad, and most people like the bad things best. Good things (for preserving permafrost, that is) include shade; lush, natural vegetation; and poor drainage. The typical homeowner would rather have plenty of sun, a well-drained, well-cropped lawn, a nice asphalt driveway, and a gravel pad for the house. The one exception to this pattern is snow. Snow is natural, but it is also an insulator at a time of year when one doesn't want a blanket on the ground. Deep snow, snow drifts, and plowed snow piles are great for small animals in that they keep the ground nice and warm, but a snow-free area surrounding the house keeps the permafrost nice and cold.

"But I *elevated* my house," you say, "and that should keep it snow-free and shady." Fine, but what about the area around the house? Once permafrost starts to thaw it is difficult to stop. Many a structure has wound up in the septic tank because the heat produced by the effluent caused the permafrost to thaw, and the expanding depression swallowed the adjacent foundation.

And what about your driveway? Keep trees and heavy vegetation on the south and west sides of the house, and put the driveway and parking area on the north.

Just for the Heavyweights

There are alternatives to elevated structures, but not, unfortunately, to cold bottoms. No warm surface can be juxtaposed to icy ground without causing the ice to melt, even if they are separated by generous layers of insulation. However, in many buildings where loads are heavy and perhaps concentrated, as in aircraft hangars and heavy equipment repair shops, elevated floors are not practical. In such cases, engineers have devised ways of keeping a floor's underside suitably cold while still placing heavy loads directly on the (frozen) subgrade.

All these techniques involve insulating the ground, then sucking heat out from below the floor. Engineers use different types of "heat suckers," including mechanical refrigeration, air ducts, and single phase and two phase thermal siphons (Figures 5a-d). Mechanical refrigeration is in essence a huge freezer with the cooling coils going under the floor. The designs are easy and positive, but the refrigerator is initially expensive and takes energy to keep it going and constant maintenance to keep it running. There are many in existence, but few have operated for extended periods of time without problems. Frequently the cooling is done by circulating the fluid through radiators outside rather than compressors in the winter. All other techniques use the concept of freezing a gravel pad under the foundation so solid in the winter that it can't thaw out all summer long. The designs are more complex, and they are critically dependent upon the weather as well as other factors involved in the design. In general, there is a floor slab underlain by several inches of board insulation over a gravel pad. The cooling system is installed below the insulation. The air duct system is exceedingly simple in that air is directed under the floor slab through ducts in the gravel, through ducts in the concrete floor or between the floor you walk on and a fake floor a few centimeters lower. The air may flow naturally, if that is enough, or it may be necessary to force it through the system with fans or a chimney. The other two systems look similar from the outside but have different thermal characteristics. In either one there is a tube that runs from below the foundation up to the air. The air cools the fluid in the tube and the ground warms it. In the single phase system the fluid is a liquid, and warm fluid is taken from under the building to the outside air, where it is cooled by natural convection or by a pump. In the two phase system, the fluid is warmed below ground and turned to a gas, which in turn is condensed in the outside part of the system to a fluid, which runs back down the pipe.

Heatstroke at Prudhoe?

There is one school of thought, prevalent in some circles, that holds we are in a period of global warming. Some say this warming is caused by hair spray depleting the ozone layer, some say our excess use of fossil fuels is raising the CO_2 level, which in turn exaggerates the greenhouse effect. Others say we are experiencing one section of the natural cycle of things, and still others say the whole issue is a hoax and we are about to enter the next ice age. In any case, the controversy serves to emphasize the fragile state of our atmosphere. If the permafrost is only a degree or so below freezing now, it may well melt naturally in the near future. Most of us do not look forward to planting palm trees in Barrow any time soon, but we are well aware of the possibility of a few degrees' temperature change in the foreseeable future.

There are Times When Piles are Just What One Wants

Piles are posts or poles placed in the ground to serve as structural supports. In tropic and temperate-zone construction they are often used but rarely visible, located as they usually are beneath the lowest columns of a bridge or building. For marine construction they are ubiquitous: most piers are built on pilings because it is comparatively cheap and quick, and because pile construction minimally obstructs the passage of waves and currents. For arctic work, pilings are popular because it is a simple way to provide thermal isolation of a heated structure and also because it is strong and permits easy passage of drifting snow or flood water (Figure 6).

Piles are installed in many ways: "pile-driving" is popular for wood, steel, or concrete piles in soft ground. What is not commonly realized

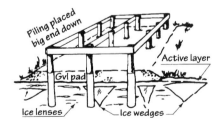

Figure 6. Typical pile foundation.

Here Eb Rice poses with a pile that came to grief in encounters with ice-cemented gravel.

is that steel piling (both open-end pipe piles and H-piles) can be successfully driven into fine-grained or peaty permafrost. Piles driven into ice-cemented gravel, however, occasionally come to grief (see following photos). Vibratory pile hammers are showing considerable promise in this area, although double-acting steam, air, or diesel powered hammers also work well.

Commonly, piles of wood or concrete have been placed in holes melted into the frozen ground by steam, hot water, or electricity (they cannot be driven into the ground). Ideally, the hole should be thawed swiftly so that its width is but little more than the diameter of the pile. The pile may then be driven into the hole with conventional pile-driving equipment. But this is not the best way: depressing failures have occurred because structures were placed on such piles before they had time to freeze back.

It is obvious, upon reflection, that a column of hot muddy water, filled or not with a pile, is a heated pocket destined to enlarge before finally freezing. If only surface-freezing has taken place, as often happens in the Arctic, a sturdy-looking, solid-feeling pile may be built upon, in the belief that it is solid. Often it isn't, and the results are disappointing. This happened many times during the construction of the DEW (Distant Early Warning) Line in northern Canada and Alaska in the early 1950s and later during construction of Inuvik, N.W.T., Canada, where the hazard was considered so real, and the freeze-back time so uncertain, that it became customary to place pilings in the fall of the year so that freeze back would be surely complete in time for spring construction.

Another consideration is unfrozen water in frozen soils. Either soils with saline pore water or very fine-grained soils exhibit this characteristic. When in doubt, check it out. Unfrozen moisture reduces the strength of the frozen soil and the load-bearing capacity of the soils.

A third and common way of installing piles is to place them in drilled holes. This method, where applicable, has nearly superseded others. A hole somewhat larger than the pile itself and deep enough to extend well below the maximum expected thaw depth, is drilled into the permafrost. The pile is then simply dropped into the hole, braced in place, and backfilled. Backfill is done in a number of ways. The most desirable way is to mix a sand and water slurry at a few degrees above freezing and vibrate it into place. Lesser, but frequently adequate, results can be attained by using the auger cuttings mixed to a slurry along with moist soils that have been loaded and compacted around the pile. Since no appreciable heat is inserted into the ground, the pile should be ready for use in a matter of days.

The holes are usually drilled by auger-type drills mounted on vehicles for easy portability and convenient setup. Such augers, if equipped with very sharp teeth, will penetrate frozen silts, sands, peat, and ice quite swiftly. (No cheap and easy way of penetrating ice-cemented gravel or bouldery ground is known to me, and in such ground the driving of piles into thawed holes will probably remain the best way for some time to come.)

The sonic drill has some potential in difficult conditions, but the equipment is not readily available, and the technique is probably not viable for home use.

How Big Should a Pile Be?

Early pile design emphasized depth and strength. Builders took the longest pile they could manage and sank it as deep as practical with the equipment available. The North is filled with case histories of projects performed similarly, and many of the resulting structures are of little use now except as fox dens.

The first modern approach was based on theory, laboratory tests, and field load tests yielding a design that had all the attributes of modern technology but still resulted in some dramatic failures. It seems that ice creeps, and therefore a foundation that would hold large loads for a short period of time would move slowly and ultimately fail over a long period of time. Current pile design takes the creep into account also, which theoretically should solve all these problems, but now engineers have discovered new worries, such as the salinity of the soil, vibratory loading, and global warming.

In general, houses are relatively light structures. Piles carry on the order of 2,000 to 4,000 kilograms each. Piles under houses are typically at least 15 centimeters in diameter. Piles with these loads and these diameters will usually hold the weight of the building if buried at least 1 meter into the permafrost north of the Brooks Range and 2 meters south of the Brooks Range. The more important issue is frost jacking and lateral support.

In the high Arctic, the piles can be short, and they act as end-bearing piles. Their depth of burial must be sufficiently far below the depth of maximum thaw so that they will give adequate lateral support (for wind or earthquake) during the thawing season.

In the sub-Arctic, though, things can be different. The active layer may be thick—up to five meters—and the permafrost is warm, weak, and willing to thaw. The task of designing a foundation in such locations is formidable. Decay may be a problem, lateral loads may become significant, and, worst of all, frost heaving may be so severe as to warrant elaborate and expensive countermeasures.

Heave, Rot, Melt, Settle, or Lean?

One way to make certain that a foundation does *not* do any of these evil things is to keep the piles frozen in place, as they would be in the high Arctic. Pile stability can be achieved through extraction of heat throughout the summer, or by so thoroughly overcooling the ground in the winter that there is not enough time in the summer to complete the thawing of the foundation area. There are several ways to accomplish this, all of which can be expensive. But sometimes a permanently frozen condition of the foundation is the surest solution, and often it is cheapest among the alternatives.

Did someone mention frost-jacking?

Frost-jacking

Frost-jacking is an insidious process that causes piles to heave upward during the cold season, with no countervailing tendency to return to the original level when thawed (Figures 7a-d).

The mechanism is simple enough, yet not widely understood. Figure 7a shows the freezing front beginning to penetrate the frost-susceptible ("heavable") soil. Any soil with more than 3% by weight of material finer than the 0.002 mm size is "heavable." Silts are worst, and a mix of clean sand and gravel is considered non-frost-susceptible. In the figure ice lenses are forming, forcing the frozen layer upward. The pile, which is frozen solidly to the heaving layer, goes up with it. This process goes on, and the pile continues to rise so long as ice layers continue to form. In Figure 7b, the heaving is complete, and in Figure 7c summer has come, thawing the upper layers. Excess water rises to the surface and is lost, and the thawed soil settles back as the thaw penetrates. By the time the thaw penetrates to the bottom of the active layer, the heaved soil has reconsolidated, and it grips the pile anew (Figure 7d). But the piles are now elevated somewhat, for until the last few centimeters of frost melted, it was held in its raised position by a frozen grip. The fence in the opposite photo is a victim of frost-jacking.

Fortunately, something can be done about it (Figure 8). If some kind of mechanical anchor is attached below the active layer, resistance to heaving can be enormously increased. Or if some method is employed to make the pile too slippery for the frozen ground to hold, the heaving forces can be reduced.

Helical Anchors and Money

Helical anchors consist of a fairly thin rod or pipe with helixes attached at several locations down the rod (Figure 4). The rod is turned and the helixes work themselves into the ground. The technology has been proven in thawed ground but should be considered an emerging

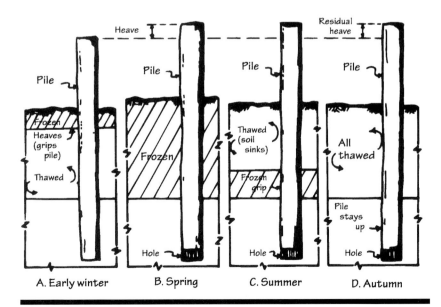

Figure 7. How frost-jacking happens.

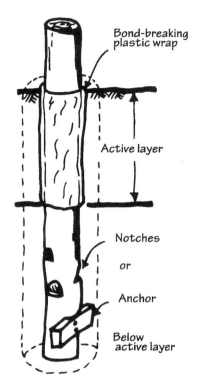

Figure 8. How to "jack-proof" piles.

Bond-breaking plastic wrap

Active layer

Notches
or
Anchor

Below active layer

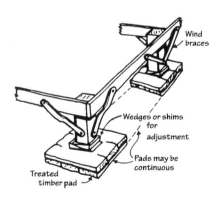

Figure 9. The post and pad footing.

Wind braces

Wedges or shims for adjustment

Pads may be continuous

Treated timber pad

technology in frozen ground. Advantages of the system include a bunch of "easies": easy installation with hand-operated equipment, light weight (easy to ship—always a consideration for Alaskans), easy to relevel with simple, factory-made jacks, and easy installation by unskilled builders in cramped quarters.

There are (of course) a few downsides that may cast a damper on an individual installation. Some soils may make installation impossible; in other soils the total load-carrying capacity might be low, requiring many anchors and more money. The lateral stability (remember those earthquakes?) may be limited if the terrain or the builder requires that all the piles be constructed vertically. Lateral loads can be improved by installing the helixes on an angle, but again, the costs rise considerably.

If You Doesn't Want Piles

There are as many options as there are Alaskans building their own houses. The vast majority, until around 1994, fell into the category of the post and pad users. A post and pad foundation is exactly what it sounds like—Figure 9 shows a typical installation. The pads may be placed directly on the tundra or on a gravel pad placed on the tundra.

The gravel pad keeps the area under the dwelling drier, so the lucky homeowner will be more comfortable when releveling the house later. It also helps limit the mosquitoes (they like standing water for egg laying). A gravel pad offers little structural value until it reaches a thickness of a foot or more. Many homeowners opt for at least this thickness, in order to keep their feet more or less dry in the soggy Arctic spring.

Wind/earthquake braces are a must, but many forget to install them, until—too late—everyone in the house is sleeping on a tilt after one good shake. The ground in permafrost areas tends to be alive; moving up with winter frost heaving and down with summer thawing. Imagine an experienced rider of elephants; the house has to be able to sway and rise, riding the ground just so.

But houses will never be as resilient as people. When things get totally out of true (and they will eventually), the dwelling must be releveled, most often with shims. Once again, there are as many ways to do this as there are Alaskans living on post and pad foundations, but Old Man Earthquake must be remembered when a builder invents a scheme. The advantages to the post and pad are simplicity, ease of construction, ease of releveling, and relatively low cost. The disadvantage is that such a foundation is never really stable.

Alternatively, one can make the foundation so stiff that it will bridge the voids and keep the structure on a plane, although not necessarily a level plane.

Children who live in plane but unlevel houses are quick to find advantages, such as playing marbles with themselves or running tricycles downhill (across the kitchen at warp speed). Parents don't seem to share their enthusiasm for the special attributes of the tilted floor, although the opportunities for good party conversation are inexhaustible, and the bar bill cheaper—it doesn't take as much vodka to get the full effect.

Steel frames and hollow concrete slab foundations have been used to construct such fun houses; the Triodetic foundation is gaining popularity.

The Triodetic foundation, a "some assembly required" three dimensional matrix of steel tubing, is typically assembled directly on the tundra by a couple of big guys using a big hammer and a lot of creative language. The dwelling is then constructed on the frame. Such a foundation becomes one with the earth in that its feet (on the high spots) sink into the tundra, while the frame bridges the low spots, leaving the bottom of the building a plane. The theory is great and the costs seem reasonable. It is, however, necessary to find one capable of supporting your structure to your satisfaction, and one size does not support all structures.

Denouement

The Site's Right

When the problems of psychology, view, radiation, drainage, drifting, flooding, convenience, and utility have been reconciled, and the designer is ready to begin his plans, and such features as roads, water supply, and waste disposal have been agreed upon, it is time to think of selecting a foundation type.

How Firm a Foundation?

In the Arctic and sub-Arctic, one can be certain that a solid foundation is possible if underlying permafrost stays frozen. This usually means that the structure be isolated from the ground by raising it to prevent heat input to the foundation soils. Once this decision is made, subsequent decisions must be made about how to make sure a cold bottom does not cause a cold floor, for a cold floor may be hazardous to your toes, and ultimately, to your structure's health.

A triodetic foundation.

Introduction

What Makes a Dwelling "Ideal"?

A tricky problem that sometimes bedevils the arctic designer is to determine what constitutes a proper house. For many people the "ideal dwelling" is identical to whatever is the fashion "stateside." If this means Ranch-Style-with-Sliding-Glass-Patio-Walls, so be it. If it means New-England-Salt-Box, that too can be done. Believe it or not, there are colonnaded Southern Colonial mansions in Anchorage and Fairbanks.

For me, an ideal house is one that combines convenience, comfort, and economy. Attaining these also requires that a house be in close harmony with the surrounding terrain and climate while doing no serious violence to the prior habits of its occupants. To provide such a structure should lead to an indigenous Arctic Architecture, a style that does not exist just yet. Nor am I prepared to invent it. I can, however, point to some of the characteristics that must surely distinguish the arctic architecture of the future, and the following comments are intended to be both practical and prophetic. For it is my belief that when the principles of handling the arctic environment are apparent, the Arctic Architecture will evolve, limited in its beauty, quality, and individuality only by the genius of its architects and engineers.

Igloo

Consider the igloo, the famous snow house of the Canadian Archipelago.[1] It would hardly meet the conditions of comfort, convenience, and durability required of a permanent dwelling. But it does have some features, derived from the harsh climate, that could well be imitated by the modern engineer or architect. Its shape—a hemisphere—not only is structurally perfect for the light, weak material used in its initial construction, but also has a minimum ratio of surface area to volume. This is exactly what is needed to conserve heat.

Not only that, but its masonry-type construction requires no false work; the whole dome can be completed by a practiced builder in less than an hour. Furthermore, within an hour or so of its completion, a snow house will sinter to a solid monolith under the combined influence of warmth and time until it is strong enough to qualify as a small fortress against both bears and storms (Figure 1).

A plan and the elevation of a typical igloo are shown in Figure 2. Several points should be noted: Entry is through a "cold trap," which contains still, cold air beneath the bubble of trapped warmth. A mere curtain of hide is sufficient to ensure that the interior climate is not greatly affected by the outside cold. The entry is placed to avoid drifting and to prevent the intrusion of the direct force of the wind. To

Figure 1. In spring, a young man's fancy... The snow for this house had to be "wind-packed" artificially. There is no arctic-type wind for the job in Fairbanks. Even so, it is strong and student Bill Kneeland, the builder, is triumphant (photo by Stu Rothman).

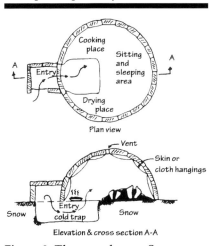

Figure 2. The snow house. Snow houses have a lot going for them.

---

1 The hemispherical igloo of snow blocks, far from being ubiquitous among Eskimo people, was unknown in Alaska, Siberia, and Greenland. It was common only among the few nomadic Eskimos of the Canadian north. Of course, in Eskimo, igloo merely means "house."

The snow house can teach us several useful things about designing the ideal Arctic house:

1. The trapped air bubble. When entrances are below floor level, comings and goings do not cause drafts or a change in indoor climate.[2]
2. The controlled vent. Totally without auxiliary power, astutely located vents control air circulation to remove unwanted heat and odors.
3. The favorable surface-to-volume-ratio. Minimizing the area exposed to the weather conserves heat and materials, both of which can be hard to come by in the North.
4. The adaptable entrance. An entrance so arranged that it can be easily adapted to varying wind conditions is a blessing fully appreciated only by those who have experienced awkward entries in windy country.
5. Material resistant to heat flow. The snow house, because it has so many things going for it (trapped heat, low surface area, controlled ventilation, ingenious doorway), can get so warm from the heat of its occupants that it might begin to melt. In fact, snow is such a good insulator that Eskimos must sometimes shave off part of the outside walls of the igloo so that it can keep cool. Modern houses, however, are unlikely to encounter the "too-warm" syndrome. This is partly because insulation costs money, but mostly because they are uncrowded: the occupants' heat production is not significant. The point is that heat can be conserved.

---

2 For long-term occupancy, some snow house users install a wood "chimney" as the air vent so that emerging warm air will not melt and enlarge the hole.

3 To use a contemporary illustration, entrance to the Boeing 737 Jetliner is through the side of the aircraft. In winter when the door is opened, the flight attendants don their bulky garb, for the cabin's warm air rushes out to be replaced by cold. In contrast, the old Boeing 727 was entered through the floor. Even when the door must remain open for a long time, cabin heat is retained and the flight attendants may greet their oncoming passengers in comfort.

accommodate changes in wind direction, snow block tunnels or walls may be appended to the house.

The "basement" through which one enters is also the quarry from which the snow for the house was taken. This procedure wastes very little snow or handling time. But the true living quarters are on the benches, the original snow surfaces, which are now elevated above the level of the "basement floor." These benches constitute the cooking and the sleeping places and are where most human activity takes place inside the dwelling. The basement floor is the deep-freeze and serves not only for storage but also to keep food waste and other debris from being offensive.

Since the snow house is so efficient as a heat trap, there is some danger—even in midwinter—that the heat of the occupants, plus that of cooking, could cause melting and eventual destruction. This is controlled by placing a vent hole near the top of the igloo. The vent can be closed at will by stuffing it with cloth or scraps of fur. Hot air can emerge through this hole, to be replaced by a gentle flow of fresh, cold air through the incompletely sealed entrance. Controlling the vent therefore controls not only the temperature but also the ventilation, minimizing both drafts and heat loss.[3]

Insulation

Everyone Knows Insulation is Important, But . . .

Myths surrounding insulation are widespread, frequently erroneous, and totally unnecessary: there is plenty of true information available to protect the prudent builder from wrong ideas. Yet I encounter daily the wildest ideas about how much insulation is "right," about what is the best kind of insulation, and about what is the best buy, so let's try to clear up the confusion.

How Much Insulation is Too Much?

The answer to these questions is more economic than technical. It depends on cost: cost of insulation, cost of construction, cost of heating, cost of money. But it also depends on climate and on the comfort level desired.

Having too little insulation can be bad. If fuel and furnaces were free, one could maintain even a tent at a habitable temperature in the coldest weather. But it wouldn't be comfortable. The walls would be frigid and frosted, and the inhabitants would find it difficult to keep things warm and unfrozen. Besides, the air would be highly stratified: too warm at the top, freezing at the bottom, and comfortable nowhere. So it is a fact that, regardless of cost, there is a minimum thickness of insulation needed for the sake of comfort.

This irreducible minimum may turn out to be too costly. In cold regions it is usually cheaper in the long run to use thicker-than-minimum insulation: a person can save more money in fuel costs than he would spend for the extra thickness. So, by selecting more insulation, one increases comfort and saves money in the process.

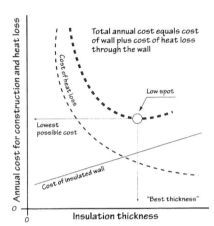

Figure 3. How insulation affects annual cost. Once you've found the low spot, you're close enough. A little thinner or a lot thicker doesn't change the annual cost much, because near the low spot, what you save on fuel you spend for insulation and vice versa.

4 Some people fall into the trap of thinking that if they already own the money and do not have to borrow, the interest rate should be zero. Not so: that money could have been invested elsewhere than in insulation, and it could have yielded income.

You Can Select the Right Thickness

Calculating the "best thickness" can be done in two ways. One way is to start with a wall design, evaluate the cost, and calculate the annual cost of heat flow through it. Adding some insulation would cost more, but it would save some heat. If the value of heat saved exceeds the minimum attractive rate of return on the invested capital, recalculate using some additional insulation, and so on until the return on the investment in insulation is no longer attractive.

Another common way is to compare annual costs for various thicknesses. Annual cost of heating is made up of the cost of fuel plus operation and maintenance of the heating system. Annual cost of a capital investment is figured by multiplying the in-place cost of the insulation in a wall by a "Capital Recovery Factor," where CRF $= i/[1-(1+i)^{-n}]$. This capital recovery factor is available from any financial handbook, and it accounts for the rental value of money (interest, i) as well as the expected life of the structure. It yields the amount calculated to pay the interest on the invested money plus enough more to pay off the investment at the end of the life of the facility.[4]

With the graph (Figure 3) in mind, some fundamental truths can be observed. Foremost, it shows that there is a "lowest point"—a "correct thickness" of insulation where the annual cost is least. Much more insulation than this, and a homebuilder would be paying too much for capital; much less than this and he'd be paying too much for fuel. The low spot represents the happy balance that ensures the lowest annual cost over the life of the structure. Having plotted such a graph for a particular project with appropriate fuel costs, life, interest rate, and so on, a builder can not only tell what the proper insulation is, all things remaining equal, but he can also do some pretty effective prophesying about "what if's."

What if, for instance, he's expecting prices to rise, generally. Should he select insulation thinner, thicker, or just the same as the graph indicates? The answer is simple once he knows how the graph was contrived: he'd select thicker insulation to take advantage both of today's lower (?) prices and the expected higher prices for fuel in the future. What if he expects that the life of the structure will be shorter than that originally figured? (Perhaps he's thinking of a heated structure for a four-year construction job, and the structure must be removed at the end of it.) In that case, fuel prices are less likely to change much over the life of the structure, and the capital charges for depreciation are very high. In such a case, a thinner insulation would be chosen.

Make the Customer Happy

Suppose he is a speculative builder and wishes to sell his house cheaply and quickly to serve a need and turn a profit. What then should he choose? I'd say he should choose what's best for his customer—and make sure the customer knows he will have the optimum thickness. Only a person who's persuaded that he is getting by for the least annual cost can be expected to pay a higher initial price. (A speculative builder should perhaps apply this consumer principle to his selection of all design alternatives.)

Moral: It is possible, for any location, for any time, to calculate what is the best (cheapest annual cost) insulation for any installation.

Naturally, the "best" could conceivably turn out to be a thickness not commercially available. The graph demonstrates that it doesn't make much difference in annual cost to change to a standard thickness—it usually happens that it is best to choose thicker, rather than thinner, insulation than the ideal optimum. By selecting "on the thick side" of the optimum thickness, the situation is improved with regard to (1) comfort, (2) slow cool-down time after heating system malfunction, (3) conservation of fuel, and (4) hedging against inflation.

What Kind of Insulation?

The available kinds of insulation include the "mass insulations," which are bars to the conduction of heat, and "reflective insulations," which fight radiative heat loss. The mass insulations in practice may be categorized as fibers or foams. Fiber insulations are made of mineral or organic "wool," and the foam insulations are made from glass or plastic, foamed with a gaseous agent to a preselected density.

For all kinds of mass insulation, the resistance to heat flow decreases as density increases, to a point. That is, the lighter the better. Most insulations useful in construction have about the same thermal conductivity. Conductivity is reciprocal to resistivity and is close to 0.25 BTU (in) per (hour)(ft^2)(F°). If Americans ever accept SI (*Systeme Internationale*) units, they will express conductivity in terms of (Watt) per (M) (C°). Unfortunately, structural strength of rigid foams is also related to density, but inversely, so that sometimes one has to decide whether to choose a good but weak insulator or a poor but strong insulator. In general, however, insulation is not self-supporting and does not need to be strong, so the lightest insulation that will hold together will do the job.

Foams, usually of glass, polystyrene, or polyurethane, are unsurpassed for certain jobs. Certainly there are no insulations so useful in contact with the ground as those made of extruded polystyrene (XPS), none so resistant to chemical attack and high temperatures as foamed glass, and nothing comparable to polyurethane for foam-in-place versatility. The thing to watch for in foams is their cost, their toxicity when burned, and their water saturability. Yet for industrial work of many kinds, foams are unexcelled and their use as insulation is sensible and increasing.

For wood-frame dwelling-house construction, however, fibrous glass insulation is the thing to use. Why? Because it uses only space available within the walls or joists, because it is far cheaper that an equivalent insulator of any other type, and because it is not smoky nor toxic when heated.

A number of people, including some dealers of building materials, believe that foamed plastic is superior to fibrous glass as insulation. Apparently this opinion comes from years of experience in a world where "you get what you pay for." And foamed plastic insulation costs, alas, four to five times as much as its thermal equivalent in fibrous insulation. Yet its thermal resistance is only slightly better. In home construction, foamed plastic should be used only where the cheaper fibrous mats will not work at all—such as underground or in damp places.[5]

From a practical point of view, fibrous insulation is the heavy-duty material to use above ground. It's by far the cheapest (in frame construction, especially) and very handy to install. Some fibrous insulations are marketed with a reflective foil backing on a thick waterproof paper membrane. An arctic builder would be wise to avoid these. Such insulation was designed to be used at lower latitudes where the optimum thickness is smaller than the usual space within a frame wall.[6] In the Arctic one does not need foil-backed insulation because (1) There is no advantage to the foil. Foil needs to have a space to "reflect into," and the designer of arctic houses will always choose to completely fill the wall with insulation because it is a better thermal barrier than an air space. And (2) the vapor seal inherent in foil-backed insulation is illusory and inadequate for a long, cold winter (although quite satisfactory for "stateside" applications). So the practical thing to do is to use thick "friction fit" batts of insulation that do not need any fasteners. Friction fit is cheaper to buy and cheaper to install. In a day or so anyone could insulate a whole house with "friction fit" batts, though it would require several days for a trained carpenter, working hard, to install equivalent foil-backed insulation rolls. In either case, a large vapor-proof blanket will be required on the warm side of the insulation to avoid the irreversible buildup of water (ice) in the wall.

The optimum thickness for wall insulation in Fairbanks, Alaska, houses at today's prices (1994), figuring about 10 percent interest rates on home loans, usually comes to 8 or 10 inches—considerably thicker than is commonly found, although many contractors aware of the State of Alaska's energy policies are building homes with thicker (as thick as 8") insulation. Much fuel and money is wasted because the walls are too thin for the best economy over the years. For roofs and ceilings, the gap between practice and calculation is even wider. Ceiling insulation thickness should approach 12 or 15 inches of fiber. Often one finds less than half these amounts, especially in homes built more than 10 years previous. The "best" ceiling thickness is greater than that for walls, of course, because ceilings can usually be insulated merely by laying more insulation in space that is already available, so that for minimum cost one can get maximum results. Walls, by contrast, have to be built thick from the start to accommodate the desired insulation. That can be expensive in both material and floor space. Hence the difference between the "best thickness" for ceilings and that for walls.

Floors, on the other hand, have a lot going for them: the external films of adsorbed air that augment the thermal resistance of insulation are at their best when heat flow is down, and the insulation itself is somewhat better in resisting downward heat flow. Also, air at floor levels tends to be somewhat cooler than air near ceilings and walls, so there is a smaller thermal gradient. When one combines these factors

5   The foamed-in-place urethanes have an additional characteristic: they are foamed with a heavy gas, which gives them phenomenal resistance to the passage of heat. When hermetically sealed, this resistance lasts a long time. Out in the open, though, or at low temperatures, most urethanes revert to about the same thermal resistance as polystyrene foam or glass fibers.

6   I'd be willing to use reflective insulation—multiple layers of foil—as insulations under a floor even in the Arctic. But not, at present, in walls and roofs. Reflective insulation is perfect, however, for space vehicles.

with the large spaces available between floor joists, one can see that insulating floors can be especially simple. Six to 10 inches in elevated floors proves to be about right, and if the floor should still seem uncomfortably cool, a carpet can disguise the fact.

If foam insulation is used, much thinner values are found to be optimum. This is not because foam is superior to fiber. Rather, it's the other way around: because foam insulation costs so much more than fiber, one can better afford to buy fuel than to purchase greater thicknesses of that expensive material. The cheapest in-place costs of foamed plastic approach 50 to 60 cents per board foot (a board foot, or "Foot Board Measure," FBM, is one square foot, one inch thick) as contrasted with eight or ten cents per FBM for common fiberglass insulation, so naturally home owners buy foam chiefly for those areas where fibers won't work well—such as against the ground.

Other insulations are sometimes tried: granulated cork, popcorn-like expanded minerals, sawdust, "reindeer moss," and others. When these are the only things available, they are fine. And some of the natural materials are plenty cheap.

It is often better, though, to use manufactured mineral wool insulation. It fits, it doesn't burn, and it costs little compared to the cost of a house. Manufactured fibrous insulation is not very susceptible to water damage, it won't settle or pack into the lower parts of a wall, and it won't rot. (I have, however, seen glass fiber insulation used as nests by some very itchy rodents.)

Vapor Barrier

Heat cannot be stopped from flowing: if a temperature gradient exists, heat will flow from a warm place toward a cooler one. Insulation can only slow, not stop, the transfer of heat. In a curious analogy, water vapor also "flows" wherever a vapor-pressure gradient exists. Vapor pressure varies more or less directly with temperature; therefore, water vapor flows in the same direction as heat does, from warmer to colder areas, unless stopped by some impermeable barrier. If no barrier exists, water vapor continues to flow "toward the cold" until it encounters a plane where the wall is as cold as the dew point. There the moisture condenses. Since in the short summers of the far north there is little opportunity to reverse the vapor flow, the collected moisture remains indefinitely, and the wall and its insulation material become even more soaked. People new to this phenomenon sometimes express surprise at finding the walls or ceilings full of ice. During occasional warm intervals in the arctic winter, some of the ice melts, and the meltwater runs into the house from a host of improbable places: lighting fixtures, window frames, baseboards. This unexpected appearance is usually attributed to a leaky roof. The common cure is heroically to shovel the snow from the roof and break the ice from the eaves. The roof, now without its insulating blanket of snow, gets cold. The remaining water in the insulation re-freezes, the drip stops, and the problem is considered cured until the next warm period, when it is typically discovered that there are real leaks this time, caused by careless shoveling during

An itchy rodent. It's not unusual to open up one's cabin in the spring and find a veritable cornucopia of dried herbs and mushrooms neatly stored in the fiberglass insulation. Unfortunately, this can occur in a home with year-round occupancy as well; somehow squirrels especially seem to delight in caching their booty in such itchy places.

Again:

1. A good vapor barrier must be placed on the warm side of the insulation. The "vapor-proof backing" on fiberglass roll insulation is not enough for the Arctic.
2. The vapor barrier should be assumed to have holes in it somewhere, so ventilation must be provided so that water trapped in the wall can escape to the outside.

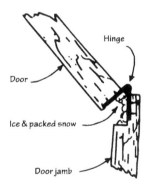

Figure 4. Enormous leverage breaks hinges or pulls the screws.

---

7 Commercial structures commonly use revolving doors to avoid unwanted drafts in winter. Don't try this in snow-drift country!

8 It is possible where the lower floor is a garage, workshop, or storage area and the upper floor the living quarters, or where there is enough space to walk beneath an elevated house.

the first go-around. It takes time and soul searching to discover that the original problem was not external leaks at all, but internal ones. The vapor barrier is either faulty or missing, and the water is that which was initially evaporated from a hundred household sources.

Two actions will cure this problem for all time: first, a builder must provide the best vapor barrier he can. This is usually a large sheet of polyethylene film. It's even better when doubled, with joints lapped generously and taped with vapor barrier tape or some other waterproof seal; it should not be punctured during the installation of electrical wiring. This barrier will always be toward the warmer side of an outer wall or ceiling. Second (and this is important), the home builder must recognize that the vapor barrier is imperfect and that some moisture will make its way through. Thus he should provide for the "venting" of all stud spaces to the outside only, so that normal "breathing" of the wall (due to atmospheric pressure and temperature change) will carry away moisture.

The water-in-the-wall syndrome will be cured if and only if these two things are done.

Doors

You Gotta Have Doors . . .

Sometimes the only ventilation in a dwelling is the air exchange of people going in or out. The syndrome is well known: a cloud of vapor-rich warm air billows up and out from the open door, and a visible surf of frigid air washes across the floor and breaks upward at the opposite wall. This can be so annoying that people feel compelled to build "storm locks." Also known as an "Arctic entry," this consists of a series of two doors that are meant to be opened only one at a time.[7]

Further, doors tend to warp, bind, frost over, and break. They tend to be torn from the hand in strong arctic winds and even wrenched from their building. This can cause consternation among the inhabitants, and gives rise to heated comment— possibly the only heat immediately available. Even if these major catastrophes are avoided, the crack along the hinge line may fill with hard-packed snow when the door is slightly ajar (jammed open, perhaps). Later, when the door is forcibly closed, the enormous leverage against the snow may crack the hinges or pull the fasteners loose. In time, the door comes off (Figure 4).

How does one avoid these problems? First, one should "trap the warm air bubble" by providing a house with entry from below. Where this is possible[8] the presence of an open door is hardly noticeable, so there is no serious urgency if that door is left open, warped, missing, or whatever. Things can be taken slow and easy, and less spirited advice will be given the person attempting repair. The second way to avoid problems is to develop a door more like the ones on refrigerators. These close over, rather than within, the door jam. This would allow hinges to be all on one side of the wall, where they would not serve as frosty cold-conductors. And it would minimize the tendency for packed snow to prevent closure. Third, if entrance cannot be through the floor, the storm lock is vital. A storm lock should be straight-through for the passage of furniture and for seeing out on a still, sunny summer day. But it might also be given a spare door in another direction to avoid the possibility of entrapment by wind or by snow drifts. The entry hall can be used, and usually is, for a lot of other things too. If large enough, it can hold boots, parkas, food, guns, skis, and game. The heated quarters open only into a cool but windless area. Again, no immediate troubles erupt if the outer door blows open. The inside door, protected, is nearly immune to problems.

. . . And Windows

Windows are costly. To me, though, they are necessary, and I like to have them facing all four directions. The best number of panes may be figured in about the same manner as the best thickness of insulation. It turns out that in most of Alaska at present costs, three panes are cheaper than two, and four panes cost about the same as two panes, on an annual basis.[9] The more spaces the heat must pass through, the better the thermal resistance. Unfortunately, glass is so costly, even when you don't consider the high costs of breakage and washing, that the most economical configuration costs from seven to ten times as much as an insulated wall, annually and forever. It should make a designer think twice before sentencing a client to the high cost of senseless windows. This is especially true at high latitudes, where the low angle of the sun makes the illuminated window even less tolerable than in climates where overhangs and foliage can offer partial shade.

Yet windows have their virtues. They allow people to see outside when there is daylight out there. They serve as emergency escape routes in case of fire or in case the doors are drifted shut. They seem to help relieve a feeling of claustrophobia that may be a part of the wilderness winter's famous "cabin fever."

For all these reasons, one should examine carefully how windows should be placed, how large they should be, and of how many panes of glass. I suggest that in addition to being large enough to escape through, with enough panes to discourage frosting-over, windows should be so situated that both seated and standing people can see out. Windows need not—should not—reach either the ceiling or the floor.

Of course, windows do tend to fog or frost over during the winter. There are many residents of the North whose view of the world is blocked from autumn until spring by frosted windows. This is almost always an unnecessary shame. Using well-understood technical considerations, one can contrive windows that do not frost, or at least not severely. It is a matter of having the inner pane sealed like a vapor barrier, together with a modest opportunity for the other spaces to breathe to the outside. Then the only condensation will be on the interior, accessible, pane. This one can be kept clear either by keeping the humidity of the air low through careful ventilation, or by blowing a stream of air against the window.

But make no mistake about it: Large windows are far too costly and wasteful to be accepted without careful scrutiny. A homeowner must

9 I'm figuring for ordinary window glass here, not the "insulating glass" that is so well advertised. It is very good (though no better than an equal number of panes of regular glass) and very expensive.

expect one's money's worth when planning windows. They're to see out of, to introduce light and air, to serve as emergency exits, and to provide a certain psychological comfort. If windows don't do these things, they are frivolous luxuries.[10]

Conclusion

The Arctic is not familiar to the ordinary designer who comes from "the outside." For example, arctic snow is not fluffy: it is usually so solidly windpacked that those who walk on it leave no tracks. The winds that pack it can often be so strong that outward-opening doors can be snatched from the grasp and whirled into the months-long night—provided, that is, that the door is openable and not totally blocked by concrete-hard drifts. "Southern" designers sometimes fail to understand this sort of thing, so they routinely follow building codes that require doors to open out.

There are hundreds of similar examples where the standard solution is inadequate or even wrong for the Arctic. In this article we have tried to pick out a few. If these remarks should add momentum to a drive for an "Arctic Architecture," I shall be delighted.

---

10 A person can, if astute, have his cake and eat it too. Strong, well-insulated shutters, placed on the outside of windows, can save a lot of heat during the times windows are not useful—that is, when it's dark outside (as in the arctic winter) or when the building is unoccupied.

Northern Construction: Heating the Ideal Arctic House

CHAPTER 4

Since I left Plumtree, down in Tennessee, it's the first time I've been warm.
—from the *Cremation of Sam McGee* by Robert Service

Save Heat

It's easy, simple, and even moral to save heat. Conservation of energy improves comfort and economy, and it reduces waste and pollution. Yet casual and careless selection of fuel and furnace, together with ignorance of insulation and ventilation, costs the world a great deal of energy every year.

People Need to be Warm

People seem to have evolved to live life in a climate of about 70°F; they feel discomfort if the temperature departs more than about 10 degrees from that. As people extended dominion and domicile over the world and moved into areas much colder than suitable for the naked ape, they carried with them the need to do something about it. That something, they discovered, was to adopt insulation—insulation of themselves by clothing and insulation of their dwelling by any means available.

Consider the Eskimos

Consider the Eskimos: with ingeniously simple garments of caribou skin, they have managed to survive for millennia in the polar climate. European explorers, encountering these remarkable people for the first time, assumed that Eskimos had evolved a physiological adaptation to extreme cold. That belief has persisted to the present, though measurements do not support it at all. Physiologists and anthropologists in recent years have asserted that Eskimos are adapted, all right, but not so much in physiology as in know-how. Eskimos have survived in the arctic environment by keeping the cold at bay with clothing or housing. The source of heat for their "70-degree climate" is their own metabolic heat, trapped by their clothing, plus a little heat incidentally derived from seal-oil lamps primarily used for cooking and lighting. Take away these things, and Eskimos suffer from the cold about as much as everyone else.

Onward, the Woodpile

As Europeans moved ever further north, they also staved off the cold, mostly by heating their cabins and by taking care to avoid going outdoors in the severest weather. By and large, that's what people still do: keep the houses warm and stay indoors, braving the elements only for short dashes to the next house or warm conveyance. That this avoidance of cold can be done so easily is a tribute to the ingenuity of our forebears who developed shrewd ways to shelter us from the harshness of the polar world.

Unfortunately the apparent effortlessness of achieving thermal comfort is often bought expensively, and irresponsibly. Old timers were highly conscious of the cost of heat: every stick of wood burned was cut, trimmed, hauled, and stacked by hand for the winter, and heat (comfort) was purchased at the cost of much effort and time—time that had to be diverted from other pursuits. Dwellings as a result were small, with few openings, and they were carefully caulked and banked.

A later generation, ours, has substituted money for work, petroleum for wood, and we are no longer so intimately conscious of the need to conserve so valuable a commodity as heat. The result is egregious waste: unnecessary production of heat with little compensating improvement of the human condition. Yet we can, with a little attention to what we're about, minimize wasted energy by using some of our neglected common sense in the selection and use of insulation, fuels, and heating systems.

Conservation of Energy

Engineers and physical scientists are acutely aware that in the production of power by burning fuel there are inescapable energy losses. The ordinary gasoline engine at best converts only about a quarter of its fuel into mechanical energy. The rest goes to produce heat. (Some of this heat is used to warm the passengers in automobiles in winter, but most has to be wasted.) Such waste is inevitable; the well-known "Carnot limit" does not allow us to convert all the fuel energy into useful work, and the 30 to 50 percent achieved by modern power plants is the best that can be done by conventional thermal cycles. Commonly, very considerable amounts of wasted heat are dissipated to the air, to rivers, or to special "cooling ponds." Under some conditions this results in thermal pollution or ice fog.

Heat "Losses" are Sometimes Recoverable

One excellent way to conserve energy, therefore, is to make use of the "waste" heat of power generation. This type of recycling is practiced frequently where power plants are constructed near cities. Here waste heat can be used conveniently to heat space; heat that is too "cool" to generate electrical power is just about right for heating. Space-heating energy, in fact, is just about the lowest "level" of energy we can make use of.

Unfortunately, thoughtless "reformers" often seek to prevent power plants from being located near cities, causing them to be so located that their low-grade waste heat cannot be used effectively. Such shortsightedness guarantees that from two thirds to three quarters of the thermal value of the fuel will be uselessly dispersed—and still more fuel will have to be consumed to heat the spaces that otherwise could have benefitted from the local power plant's waste.[1]

Total Energy Systems

In the case of certain buildings, even in lower latitudes, thoughtful engineers have taken advantage of suitable ratios of power, heating, and cooling needs and have provided a "total-energy package," or

1 We conservationists should be concerned that our recommendations (such as, "Get the power plants moved out of town!") do not lead to evils worse than the ones we fight.

ogeneration system, a self-contained system that so efficiently avoids
energy waste that it can compete with the local electric power utility.
These systems sometimes are so "total" that the energy in the building's trash, burned to augment regular fuel, provides a sizeable fraction
of the total heat and power need. In addition, the export of garbage,
sewage, and trash is reduced. Such systems are not available off the
shelf just yet, but there are many being used, and more are being designed by engineers who wish to serve (and prosper) by solving at once
the demands of economics, conservation, pollution, and independence.[2]

About Fuels

So far as heating homes, villages, and cities is concerned, the fuel
available is pretty much limited to wood, coal, natural gas, and several
grades of oil.[3] Electricity is not a fuel so much as it is a means of energy
distribution. (Nevertheless, it is possible to "build an electric fire" in an
otherwise ordinary stove or furnace.)

Electricity

For simplicity, ease of installation, and safety, electric heaters are unexcelled. Electric heat costs a lot, however, compared to heat from other
sources (about twice as costly as oil per BTU delivered and over four
times the cost of coal—see Figure 1). Worse, it is usually produced in
inescapably inefficient fuel-fired power plants. This means that between
two-thirds and three quarters of the fuel's heat value has been wasted
already before the electricity passes through the wires. In contrast,
only about one fourth of the heat from direct burning of fuel is lost "up
the stack" in stoves and furnaces (and some of the new high-efficiency
gas and oil furnaces actually recover for heating over 90% of the heat
generated during combustion). Therefore, electric heat uses about three
times as much fuel as does heating by furnace. Perhaps one should aim
to avoid the use of electric heat except where it is clearly so convenient
that it justifies the extra environmental and monetary costs, such as
when cooking and ironing.[4]

One of the great attractions of electricity is the appearance of waste-
lessness. At the point of use, there is no smoke, no dirt, and no furnace
room. Moreover, you get every kilowatt-hour you pay for. To add to
these merits, the initial installation cost is usually lower than for other
central-heating systems, and it is easily accommodated to the dictates of
a heating system, so long as the power source is dependable.

In further defense of this remarkable (however wasteful) heating
method, it can be fairly said that, where the only practicable source
of energy is fuel unsuited to domestic use (as with low-grade coal or
nuclear fuel), electricity may be the only reasonable way to heat.

Wood

Wood has the considerable advantage of being renewable. If care is
taken to harvest no more than is grown each year, and if the wood to be
burned is more valuable as energy than for some alternative (lumber,
paper, plastic), then wood serves well. In all the forested regions of the

Service, and North Pole Gravel for
supplying these updated figures for
2007).

2 May their designs prosper. May their
hardware never fail...

3 Liquified gas (LPG, LNG, propane,
butane, and so on) is rarely competitive
with heating oil for space heating. Like
electricity, though, it is great for cooking
and lighting.

4 Electricity has long been generated
at efficiencies higher than 85% from
"hydro plants," where falling water—a
renewable form of solar energy—is available. Electric heat from such sources is
environmentally acceptable, I should
think. But to me, it is esthetically (and
morally?) improper: it is a bit like grind-
ing up tenderloin for hamburgers. It is
conceivable that, when an atomic fusion
reaction becomes available for producing electricity, mankind will be so rich
in power that we can again contemplate
using electricity for house heating.

Figure 2a. Yukon stove: generalized, folding "Model T."

North, wood is virtually the only fuel considered by rural people for heating. Its supply is assured, its acquisition requires no cash money, and its use rarely requires sophisticated (electric) control systems. Individual cabins, heated by simple wood burning stoves, have been and remain a popular style.

Even industrialized areas sometimes make use of wood for heat. Slash, mill waste, bark, and other wood wastes have been used either directly or as "hogged fuel" (chipped wood) for both heating and power production. The use of sawdust for house-heating has even caused the development of an elaborate technology of sawdust burners and their control systems. The sawdust furnace, however, has lost popularity even in the wooded areas of the Pacific Northwest. Cheap oil and the rising value of sawdust as a raw chemical have contributed to its disappearance. Sawdust burners are an endangered species and will require a very special environment if they are to avoid extinction.

The use of wood for fuel, however, will linger long. Not only is wood an environmentally acceptable fuel (burning of waste wood may be its ultimate disposal method) but also may be an aesthetic coup. The burning of wood in open fireplaces has such an appeal to the human spirit that large wood-products companies find it profitable to process wastes into fireplace "logs."[5] Such logs, intended solely for the luxury market, are far too expensive for house heating.

Wood for fuel, despite its many qualities (pleasant aroma, renewable resource, clean handling)... has lost its place to the more convenient oil and gas, and is unlikely to regain popularity except among forest dwellers who have time and chainsaw oil on their hands.

Oil—the Vanishing Choice...

For most camps and villages in the North, especially those above the tree line, oil is the only practicable fuel to deliver, store and use.[6] The lighter grades of oil (Number-1 and Number-2 stove oil) are used in home-heating systems, while heavier grades ("Bunker C," "Heavy X"), which are cheaper and contain more energy per pound, are used in major heating and power plants. Very heavy oils are usually the residual oils left after the lighter liquids (naphtha, gasoline, jet-, diesel-, and stove-oils) have been distilled out of crude oil. They usually require heating to be pumped and handled.

Because of the convenience of handling and transporting oil, it has largely supplanted wood and coal as a favored fuel for space heating. That's one of the reasons the world is now so desperately seeking new supplies of this exhaustible resource.

Natural Gas

Natural gas, like oil, is so valuable and so scarce that its price, too, must rise. Other than that, it has nearly all the virtues of oil for heat, plus a capability for use as cooking fuel. Recent gas discoveries around the Arctic rim assure that gas will become a strong competitor with oil as a space-heating energy source.[7]

5 It is a commentary on our technology, wisdom, or folly that ersatz fire logs made in Idaho are available—and are bought—in wooded Alaska.

6 Barrow, on the North Slope, has natural gas piped to its houses. Where available, such fuel is usually competitive. In Barrow's case, it is more than just competitive. Government-supplied, it is also government supported and is vastly cheaper than any other fuel.

7 Our descendents, when thinking of our era, will regard us as barbarians to have used so much of the world's irreplaceable oil and gas for such low-grade purposes as heating. Oil will, in time, be far too valuable as a raw material to permit its use for heating. We'll use waste energy, waste material, and coal for that, by then. Eventually, even coal will be used primarily for its chemicals, not for its heat.

Figure 2b. Yukon stove from fuel drum: "Model A."

Figure 2c. Yukon stove modified for use as a cook stove.

---

8 The secret of success in using Yukon stoves is to have plenty of wood cut in advance, plus spare sheet metal for patches.

9 "Master" is a trade name. There are several other makes of similar heaters available as well. There is no generic term other than "portable space heater."

Coal: the Black Sheep, the White Hope

Coal is beginning to win back the popularity it formerly enjoyed, before petroleum became so cheaply and so universally available. It is much cheaper than oil per delivered BTU, but its use must be carefully contrived to avoid problems with coal dust, ashes, cinders, and air pollution. At present prices, householders willingly pay twice as much for fuel oil to avoid dealing with the annoyances of a coal-fired furnace. All the same, the world has far more extensive reserves of coal than it has of oil, and even householders must one day acknowledge that.

Stoves and Furnaces

The Yukon Stove

The "Yukon stove" is famed throughout the northern forest of North America. There are countless designs: some are "store-bought" collapsible models, and some are even equipped with a kind of light "grate" to encourage the flow of air to the fire. Probably the secret to success when making Yukon stoves is to have a long chimney, for good draft.[8]

Such stoves are fine for locations where wood is the logical fuel, but they are not so well adapted to coal or oil. For these fuels, factory-built stoves are favored. Figures 2a, b, c, and d show the evolution of the basic Yukon stove.

The "Master" Heater[9]

Builders and others who need dependable, temporary, portable heat sources commonly select a gun-type oil burner that needs no flue. The typical burner of this sort uses electricity to run the fuel pump and the blower. Fuel is contained in a built-in tank. Such a machine is fully portable within range of an electric power source, and can be bought in many sizes, some of which produce considerably more heat than an ordinary residential furnace.

Having no chimney, these heaters are not used for permanent installation in dwellings. But their exhaust emissions are fairly clean. Virtually 100% of the heating energy of the fuel is turned into useful heat (nothing lost up the stack—no stack). The end-products of fuel combustion, of course, are carbon dioxide and water. (About a gallon of water is made for each gallon of stove oil consumed. Of course it is produced in vapor form, and is not viable except where it appears as frost or dew on cold windows and eyeglasses.) The high moisture content, plus the odor of the heated gases, make these useful heaters unsuited to living quarters; they are used for temporary heat at construction sites, warehouses, and wherever fast emergency heat is needed.

Stoves

Stoves are used for small spaces, and, if fueled by oil or gas, they can be fairly automatic in their operation—at least in the sense that they do not require constant attention. Oil and gas stoves have a control device (carburetor) that allows the householder to meter the fuel and to vary the heat according to need. Such stoves can even be equipped with a

Figure 2d. Yukon stove modified for space heating.

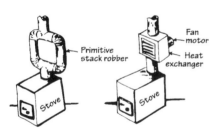

Figure 3. Two different styles of stack robbers. When draft is ample or excessive, stack robbers are used to extract heat that otherwise would be lost to "warm up the whole outdoors."

10 Heat distributed from a single source provides "central heating." An English friend of mine, apologizing for the fabled discomfort of English houses in winter, observed, "We British could not invent central heating, so we invented tweed."

thermostatically controlled valve to turn the flame up or down without human help.

The "Stack Robber"

When the draft up the chimney is strong (that is, when the chimney is tall or when the weather is very cold), it is possible to increase the efficiency of a stove or furnace somewhat by taking heat from the stack gases. A heat exchanger ("stack robber") is placed in the flue to extract some of the excess heat and restore it to the interior where it belongs. Stack robbers range from the very simple to the quite sophisticated ones designed to maintain constant draft and therefore high furnace efficiency (see Figure 3).

Furnaces

When buildings become larger than one or two rooms, furnaces supplant stoves for heat. A furnace, perhaps housed in its own furnace room, produces heat destined to be distributed throughout the structure. Two main methods are used for heat distribution: forced-air and hot water. For large installations, steam is popular.[10]

All furnaces (sometimes called "boilers" in the case of water or steam-heat systems) have a means of burning the fuel in a chamber that is usually lined with fire brick. In water or steam systems, the flaming gases are passed around tubes filled with water, which becomes hot and pressurized, ready for export through the distribution pipes. In air systems the heat is transferred directly to air in a "jacket" around the fire box, then transported through air ducts. Furnaces work under either a forced air (air moved by fans) or a gravity air system.

Coal-burning furnaces or boilers are usually fed by an automatic stoker of some kind. In a typical system, coal about the texture of coarse sand or fine gravel is conducted through a pipe or channel by an electric-powered screw that forces the coal into the firebox, where it is consumed in a bed of burning coals through which air is forced. When the furnace is heating, the screw and the blower work simultaneously. Between intermittent periods of operation, the coals remain glowing, ready to resume combustion at the next call for heat. Daily removal of cinders (clinkers) is necessary and so is an occasional discharge of ash. Otherwise, the operation is automatic.

Have Gun: Will Heat

Almost all oil furnaces use a gun-type burner. This burner has evolved into a simple device designed to be totally automatic. It shoots a mixture of fuel and air into the firebox, where the mixture burns as a plasma of flaming gas.

The "gun" consists of an electric motor, a fuel pump, an air blower, a sparking device, and a nozzle. Fuel is pumped under high pressure through the tiny nozzle. The resultant spray is a fan- or cone-shaped mist that is easily ignited by an electric arc blown into the oil mist by using the same air stream that provides oxygen for burning. (Upon starting, these ingenious devices sense whether ignition has taken place. If not, the furnace is automatically shut down. The turn-on cycle is repeated a time or two, and if the fire still does not start, the unit remains shut down. Human intervention is then required for operation to resume.)

Controls

Thermostats—Basic to Control

One of the most remarkable aspects of modern heating systems is that, when electric power is available, the systems work with almost no attention from the householder.[11] This, of course, is because of the cleverness of control circuits and the dependability of their design.

The simplest of systems uses a thermostat that senses whether room temperature is high enough. If not, the thermostat closes a switch that starts the furnace. When the temperature rises to the desired level, the switch is opened and the furnace rests. Where one thermostat can serve to control an entire house, there is no need for additional equipment.

If zones are desired, however, each with its individual requirements for heating, a slightly more elaborate system is called for. Zone controls could consist of a separate furnace for each zone. Far cheaper for ordinary dwellings, however, is a system whereby only separate pumps are used for each zone. The zone thermostat turns on a circulating pump (or air blower) for its zone only. Of course, this requires yet another loop: when the boiler water is cooled (a pump having transported heat away to a zone), a thermostat (called an aquastat) within the boiler water closes a switch to start the furnace. (Air systems work in about the same way.) The pump motor and the furnace motor are thus coupled only by the temperature of the liquid common to both. The one sends heat to the correct zone, and the other maintains a ready reservoir of heat.

To save even more hardware (and money), zone-control valves are used. In such a system, a lone circulating pump can accommodate several zones. Room temperature is sensed by a thermostat that opens its control valve. When the valve opens, it closes a switch that starts the circulating-pump motor. Therefore, the pump will be running any time any one of its zone control valves is open—that is, any time a thermostat calls for heat. The furnace itself will come on only when circulation through cool rooms has lowered the temperature of the boiler water (Figure 4).[12]

Beware the Super-clever Design

Greater complexity is often built-in where dwellings are particularly luxurious and the heating system designer wishes to display his virtuosity. Devices that sense the outside temperature, the amount of incoming or outgoing radiation, or the wind effects are often coupled to the control system to anticipate a need and to provide for more or less heat before the interior of the building can have time to respond. And similar arrangements can be made to accommodate heating and ventilating to crowds, fireplaces, and open windows.[13] There seems to be no limit to the levels of ingeniously contrived control loops. Sophistication has its price, however: operation of the more complex systems requires frequent attention by a service technician, who needs to be at least as clever as the designer to be able to figure out the intricacies of the

11 It never ceases to amaze me that seemingly complex home heating systems can work for so many years with so little servicing.

12 Of course, thermostats do not ordinarily operate line switches. Operating, as they do, on low voltage and small wires, they energize relays that operate the switches. The effect is the same as if the thermostats were also the switches. But the cost is less, the reliability more.

13 Controls need not only be electric. Pneumatic and hydraulic controls are also used.

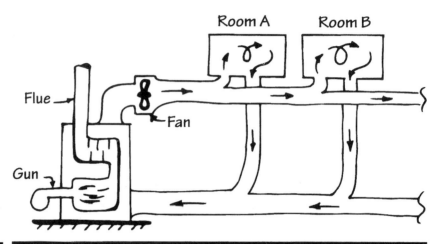

Figure 4a. Hot water heating system. Circulating water is generally kept at a fixed temperature in the boiler. When any zone calls for heat, the zone control valve opens and the circulating pump starts. Water for domestic use is heated in separate coils and does not mix with the water circulated.

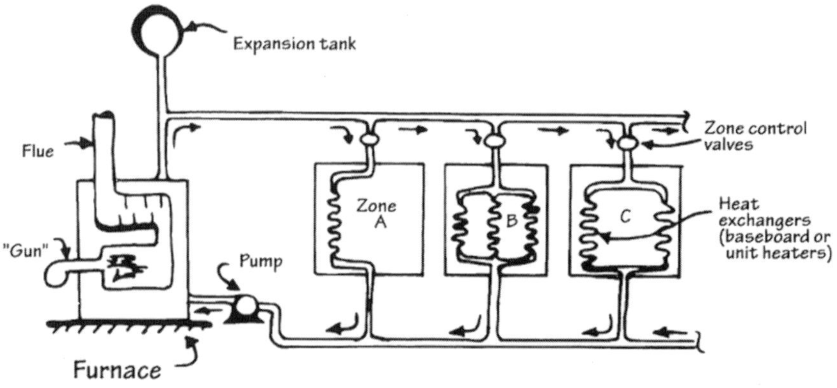

Figure 4b. Forced air heating system. Rooms can be "balanced" manually by adjusting valves (dampers) in ducts to each room. Elaborate systems may use thermostat-controlled zone control valves.

design. For the technician must not only adjust the system according to the original plan, but must also discover and compensate for troubles the designer did not anticipate. This situation can be particularly vexing when various stray interactions in the control system cause the mechanical equivalent of a nervous breakdown. Good engineering frequently means in simplicity with provision for manual operation if ever the (simple) system is undergoing psychoanalysis. For it can well be that there is no one in the vicinity who can give the system enough sympathy, understanding, discipline, and T.L.C. to keep it working properly.

Ventilation

The properly built arctic house is equipped with tight walls and roofs, each rendered impermeable by a vapor barrier. The windows are sealed, the doors carefully weatherstripped. The occupants, encouraged by inclination and by their companions, avoid unnecessarily opening these tight doors and windows.

The result can be disastrous. Hardly a year passes in Alaska without a report of some tragedy caused by lack of ventilation. All fuel burning requires oxygen, of course, and in the coldest weather high fuel consumption demands astonishing quantities of air to support combustion. A properly tight house could therefore have too few drafts to provide

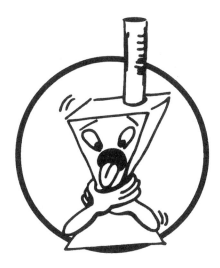

For heaven's sake, let your furnace breathe.

enough air for even the furnace, not to mention the occupants. Even if the stove or furnace were able to continue efficient operation when air-starved, the people would be in competition with the stove for oxygen, and could suffer symptoms of anoxia. (One such symptom is that the victim is unable to recognize his peril.) But the stove does not operate properly under such conditions, and with diabolical resentment may even produce carbon monoxide. The result could be, and too often is, death. House trailer and recreational vehicles seem particularly susceptible to the "air starvation syndrome." Houses are slightly less vulnerable, perhaps because their under-ventilated furnaces can misfire in such a way as to ignite the structure and destroy the evidence. The cure, of course, is to provide unclosable vents to ensure that the stoves—and the inhabitants—get enough air to behave normally. There can be no excuse for failure to provide ventilation. For the ordinary dwelling, even in the coldest weather, a vent six inches or so in diameter, so arranged that it cannot be clogged with snow or trash, is enough to provide the furnace with air.

Heat Distribution

Steam

Steam-heat systems are rarely used for individual houses but remain common for large complexes, apartment houses, campuses, and institutions. Steam's great merit is the economy with which a lot of energy can be transmitted through small pipes. A disadvantage is that steam boilers tend to be strong and heavy and need frequent attendance.

Insulated pipes carry steam to distant areas where it is introduced into heat exchangers (radiators). The heat exchangers condense the steam into water and release its considerable heat-of-vaporization. The heat exchanger may be equipped with a fan controlled by a thermostat, or it may be a simple radiator whose heat is transmitted into the room by radiation or by gravity-induced convection. Once the system is balanced by adjusting valves in the pipes, little further control is necessary—the temperature of the room itself governs the rate of condensation and therefore the amount of steam received from the boiler. Water from the condensed steam (the condensate) returns to the boiler to be recycled into steam. In some small systems, the steam pipes themselves serve as condensate return lines, so they must slope downward toward the boiler in all areas. Otherwise, separate condensate pipes are required. Careful design is required for extensive steam systems, for numerous frailties are possible that result in noises, freeze-ups, and trouble.

Hot Water

Residences and small institutions commonly use hot water for heat distribution. Water is circulated through the living spaces, returning continually to the boiler to be reheated. As with steam heat, the energy is transferred to the rooms by heat exchangers. Baseboard units, heat exchangers consisting of finned tubes shrouded with light sheet metal jackets to encourage convection, are often placed along the base of exterior walls. Most houses provided with hot water heat use these exclusively, although for larger installations fan-driven heat exchangers are also available. The advantages of a hot-water system are simplicity, controllability, and reliability. A disadvantage is its vulnerability to freezing in the event of a long-continued power outage.[14]

Forced Air

Systems that depend on the forced convection of air are common, effective, and somewhat cheaper to buy than steam and hot-water systems. The furnace is dry, and need not be built to withstand high pressures. A simple sheet-metal jacket surrounds the (finned) firebox. Air circulating within the jacket picks up heat and is then conducted through light air pipes (ducts) to the spaces to be heated. Control is by fans and/or valves (dampers) in the ducts. Return-air ducts are used to cycle the air to the furnace for reheating.

Forced-air heating lends itself to controlled ventilation, filtering, and humidifying. The advantage is forced air can't freeze. Disadvantages, which tend to be minor and avoidable, are noise, space, and drafts.

Often, forced-air systems are used in combination with steam, water, or electric heat to grant flexibility of design and operation, because air systems lend themselves to ventilation, vapor control, and air filtering.

Gravity-air Systems

It isn't always necessary to use forced air circulation. In two-story houses, gravity alone is adequate to circulate the warm air because the ascending ducts act like chimneys. As the warm air grows cooler and heavier in the distant rooms, it returns to the furnace in return-air pipes. Control is automatic and without special devices. Once the system is balanced, circulation waxes and wanes in accordance with the demand on the system. When the spaces call for heat, the thermostat starts the furnace. The extra heat induces a more vigorous circulation, which slows as the rooms reach their set temperature.

Electric Heat

Simplest of all systems to design and control are electric ones. There are numerous kinds of heaters available, from simple "hair dryer" type units that blow air through a heated grid, to baseboard units, infrared radiant panels, and heat-lamps. Any of these can be arranged to respond to either manual or thermostatic control.

Besides their simplicity and low installation costs, the radiant types of heaters are unexcelled for giving immediate heat; the occupant of a newly heated room receives warmth long before the air has fully warmed, whenever he is illuminated by the radiation.

All systems, electric or not, make use of a certain amount of electric heat, of course. Lights and appliances give off appreciable heat in addition to their other duties, and this, of course, helps keep the house warm. (If your house is heated electrically, you need not worry that your refrigerator is inefficient and has to run a lot; you're making use

---

14 Antifreeze in the boiler water reduces this hazard considerably.

Within the next twenty-five to thirty years, we might expect our heating priorities to sound something like this:

1. It's best to use the most economical thickness of insulation, or thicker.
2. It's better to use waste heat rather than fuel whenever possible.
3. It's better to use waste fuel (wood scraps, sawdust, trash) than to use fossil fuel, if you can.
4. It's better to use coal than to use oil or gas (we've bigger reserves of coal, and, so far, fewer uses).
5. Electricity should be used for heat only when nothing else will serve (cooking, ironing, etc.). It is too valuable to use as mere space heat.

of all its waste heat. If your house is heated by nonelectric means, the incidental heat still is useful; it is merely more expensive.)

Conclusion

Heating systems of whatever kind are users of low-grade energy—energy of a type easily conserved by attention to insulation. When such energy is needlessly dispersed (increasing the entropy of the universe), it is lost forever to man's use. Higher-grade energy, the kind convertible to mechanical work, degenerates eventually to heat. It therefore seems wise when designing heating systems to seek ways to (1) trap heat by carefully and thoroughly insulating heated spaces, and (2) to divert as much as possible of the waste energy (from engines, generators, etc.) to heating uses.

When trying to trap heat, we should note that even if insulation were perfect (which it cannot be—all heat barriers are leaky) there would still be considerable heat expended in warming up cold air that later is exhausted as waste. In a well-insulated structure only about half of the heat is lost through the walls and ceiling. The remainder is expended largely in heating the air needed for ventilation. Thoughtful engineers have developed ways to save a share of this heat, and these devices, called air-to-air heat exchangers or heat recovery ventilators, (HRVs) are now available to the householder and can reclaim up to 60 or 70% of the heat discharged in exhaust ventilation air.

When trying to divert energy, one normally thinks of using waste heat from power generation for space heating in winter. In summer it can be diverted to heat soils, greenhouses, or even swimming pools. There is little difficulty in finding a use for heat in the Arctic (yet we now, in true temperate-zone fashion, throw a lot of it away). It is also possible, and conceivably economical, to store heat in summer for later retrieval and use.

Heat pumps are ingenious systems designed to extract, from low-temperature heat sources, a modest quantity of energy at higher temperatures. A power plant cooling water, while perhaps too cool to supply adequate space heating directly, can serve as a source of energy for a heat pump (reverse-cycle refrigeration unit). This process consumes some electrical energy, but nowhere near as much as would electric heat.

In the future, we will see economic incentives to conserve heat. Liberal use of insulation is imperative to conserve our dwindling resources. What's more, it increases our comfort and reduces our problems with moisture in buildings. Coupled with the saving of heat now commonly wasted, we note with appreciation that insulation saves resources, saves money, saves problems, and increases comfort. It's nice to know that the moral thing to do is also the cheap way to go.

Northern Construction: Vapor Barriers

CHAPTER 5

Introduction

Water vapor causes problems for everyone—and most of us don't even know it. We know we have problems, all right, but we blame them on some other cause, some other mechanism. For instance, when we get a heavy snowfall in December and water starts dripping through the ceiling, we diagnose the cause as a leaky roof and blame the material or the builder. Or when the base of the wall turns dark and damp during a January thaw, and water seeps out to wet the kitchen floor, we suspect bad plumbing. Or when carefully fitted multipane windows frost over in October and don't regain transparency until March, we sigh, shrug, and conclude that "such is life in the Far North" and that nothing can be done. When the outside walls become diseased with peeling paint, or when the tarred roof blisters and breaks, we are quick to blame the paint, the painter, or the roofer. Always we know these things will happen, and we are confident that we know why.

We are wrong. Often.

In each case, the villain is probably water vapor—or, more properly, the villain is a designer not yet accustomed to designing for long, cold winters; his *agent* is water vapor.

Man—The Tropic Animal

Our trouble begins with the habit of living in a tropic environment, regardless of the local climate. It is fundamental that warm air can hold many times more water cold air; when the heated air of our living enclosures is surrounded by a cold climate, we expect a few local disturbances, even if we are unsure of just what they might be (Figure 1).

Air at 70°F, when saturated, holds about 0.016 pounds of water in each pound of dry air, whereas air at 10°F can hold at most 0.0013 lbs, and the amount diminishes still further as the temperature descends. The effect of this is to cause dwellers of the North to live in a very dry environment indeed: below-freezing air, even if originally saturated, will have humidity of 25% or below when warmed to house temperature.

This percentage can be raised for comfort and health by installing a humidifier, but this purchase is rarely necessary. Cooking, washing, breathing, and the drying of clothes all add to the moisture of the warm air. According to the ASHRAE Handbook of Fundamentals, a family of four may produce as much as 25 pounds of vapor daily without clothes-drying (319). But not all this water remains in the air: normal ventilation removes some; some forms dew or frost on windows, doorknobs, and other cold surfaces; and some, alas, condenses unseen in the walls and roofs. There is an upper limit to the amount of moisture the air in a room can hold, determined by the temperature of the coldest surfaces.

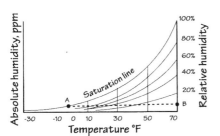

Figure 1. Humidity versus temperature. Saturated air (100% relative humidity) at 0°F when warmed to 70°F without addition of water will have a relative humidity of less than 10% (Line A-B).

If a window, for instance, is colder than the dewpoint temperature for the air, water vapor must condense as dew or frost. This dries the air somewhat. If higher humidities are desired, a continuing supply of new moisture must be added to feed the thickening condensation. For practical reasons, therefore, a room has a limiting relative humidity. Higher humidities than the "limit" are hard to maintain. But even humidities well below the limit for visible condensation still are capable of filling insulated spaces with frost or dew.[1]

Whence Cometh the Flood?

If we added no moisture at all to the heated air, its relative humidity would be low, low. But we cook. We bathe, we wash, we do laundry, we carry in snow on our boots. We breathe. All these things add moisture to our air, and perhaps that's a lucky thing, for otherwise the humidity would be so incredibly low that our health might suffer.[2]

An outstanding source of humidity is the burning of hydrocarbon fuel. Cooking with gas—that is, heating over a direct flame—produces great quantities of moisture. As a rule of thumb, burning a pound of gasoline, propane, or light oil will produce, about a pound of water. This is a common way to fog windows and eyeglasses and to saturate walls. During winter construction, every effort is made to keep work sites warm, and builders commonly use unvented "Master" heaters to do the job. For every 350 gallons of fuel thus consumed, about a ton of water is produced. Some of this is carried away by ventilation; but unless the designer is on his toes, much will condense within the walls and roofs.

How Does Vapor Migrate?

As a reasonable approximation, vapor transport by diffusion through materials follows this relation:

$$w = \mu \, \frac{dp}{dx}$$

where

w = weight of water vapor per area per time

μ = permeability

$\dfrac{dp}{dx}$ = vapor pressure gradient

This equation is a form of Fick's Law, and it is analogous to the Fourier's equation used in computing the flow of heat. Unfortunately, μ is not a simple constant. It varies with relative humidity and temperature in a complex fashion. Using $\overline{\mu}$ as the average through a material, the equation becomes:

$$w = \overline{\mu} \, \frac{(p_1 - p_2)}{x}$$

where $p_1 - p_2$ is the vapor pressure difference across the element, and x is the length of the vapor flow path (i.e., the thickness of the material).

For many calculations it is convenient to use *permeance*, M:

$$M = \frac{\overline{\mu}}{x}$$

1 Grease has a vapor pressure, too. In cooking, and especially in baking, grease vapor diffuses through the air. The "dew point" for grease and oil is quite high, so condensation occurs mostly on visible wall and ceiling surfaces.

2 Some doctors claim that arctic indoor humidity is too low at best, and they prescribe humidifiers. They do not always recognize that they inadvertently may have prescribed fogged-over windows and wet walls. Later they may ponder how to treat "cabin fever" and "flat wallet."

where M is usually expressed in *perms*, whose units are 1 grain per (hour) (ft²) (inch of Hg vapor pressure difference). Resistance of a sheet to vapor flow is $1/M$, called a REP. The total permeance of an assembly is:

$$M_{total} = \cfrac{1}{\left(\cfrac{1}{M_1} + \cfrac{1}{M_2} + \dots \cfrac{1}{M_n} \right)}$$

This quantity can be used in a modest permutation of the basic equation:

$$W = M_{total} \, \Delta p$$

where:
W = the rate of water in grains per hour per square foot.
Δp = vapor pressure difference, inches of Hg.

Although diffusion of vapor in air follows Fick's Law pretty closely, the permeability of air is quite high so the resistance of air films often may be ignored in practical calculations. (This circumstance is in strong contrast to the resistance of air films to heat flow. Considering heat flow through windows, for example, the air films are the *dominant* resistors. Not so for vapor flow.)

An Example Problem

Imagine a more or less typical wood frame wall of the simplest sort used for Arctic construction: 2" x 6" studding, covered with unpainted exterior plywood and insulated with fibrous glass (Figure 2). The warm side air temperature is 65°F with a relative humidity of, say, 50%; on the cold side the temperature is –35°F with a moderate wind and a relative humidity of 100%.

Using the relationships developed above, it is possible to calculate for a vapor pressure gradient consistent with the permeance of this wall and its parts, shown as line A. This line is something of a fiction since it is based on the proposition that there exists a steady-state passage of vapor clear through the wall, each portion of which contributes to the total resistance to vapor flow. But this cannot be. Line S is a graph showing the vapor pressure in air at saturation for the temperature that exists in the wall. Lines S and A cross, about a third of the way into the wall. This seems to imply an inconsistency: steady flow of water vapor requires that vapor pressure on the cold side of the crossing, C, be higher than is possible. In fact, at C the air is saturated, and at any position to the right of it, the air is saturated and has a much lower vapor pressure (and therefore has a much lower water content) than at C. The "missing" water has reached its dew point at C, and condenses out as dew or ice in the region from C outward. (It actually accumulates on the inside surface of the outer plywood sheathing.)

In this case, the vapor flow into the wall could be as great as 0.5 grains of water per square foot per hour but could be only a few thousandths of that quantity out of the wall. Even if the material did not soon "load up" with ice and totally block the vapor, there would still be a continual increase of moisture in the wall; the available vapor

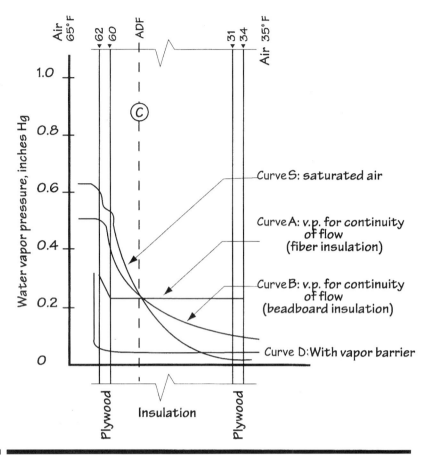

Figure 2. Water vapor flow through building materials. Curve S for "saturation vapor pressure" shows the maximum vapor pressures possible for the temperatures encountered. Where this curve is below curves A, B, or D, there is insufficient vapor pressure available to drive vapor to the right. Even with a fairly good barrier (curve D), there will be some condensation on the inside of the outer plywood.

pressure gradient through the outer portions of the wall (along line S) is insufficient to cause significant outflow. If the relative humidity on the warm side of the wall were reduced to, say, 20%, all else equal, the rate of increase of moisture in the wall would decrease to about 40% of the previous amount, but there would remain a buildup of moisture in the wall, with all the attendant circumstances appertaining thereto. Research (McFadden 1988) has shown that a moisture content of as little as 3% by volume can degrade the insulation effectiveness to one-third its original value. I am inclined to believe that, and to believe further that saturation of insulation by frost or ice reduces it to a thermal conductance approaching that of masonry, which is very poor indeed—and there certainly are many cases where walls have been cut through in the winter to reveal a "concrete" of fiber-filled ice within.

Minus 35°F, however, is perhaps not a typical temperature, even in the Arctic, so maybe the point is overstated. Nevertheless, a similar calculation for any other reasonably typical wintertime temperature reveals the same story: walls fill with moisture because it can enter much more readily than it can depart whenever colder temperatures prevail. When the outside temperature warms somewhat, some of the trapped ice melts, and the liquid water makes itself apparent as the walls leak or the ceilings drip. Worse, the water can also perform its mischief secretly by inviting unseen mildew, rot, delamination of plywood, deterioration of paint, ad infinitum.

The calculations reveal important truths about moisture buildup, and they point the ways by which something can be done about it. They can even be refined a lot to reveal the effect of diffusion of vapor through the adjacent air and the influence of the changing permeability of the material with temperature.

Sad to say, such calculations mostly reveal the scientific virtuosity of the calculator, for the important mechanism of moisture buildup is not diffusion at all, but is something infinitely less susceptible to accurate calculation.

To illustrate, let us picture a double pane window (Figure 3). Here there can be little question of permeability, for glass is totally impervious, and its frame is not significantly permeable. Why, then, does one often encounter a buildup of frost on the inner face of the outer pane? The answer lies in the "breathing" of the space. As the outside temperature fluctuates, as the sun illuminates the window, as the barometric pressure changes, the air in the space between the windows expands or contracts. Unless the space is sealed bottle-tight, some air must enter

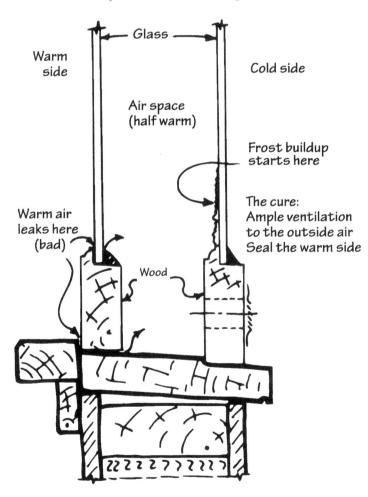

Figure 3. Multiple-pane windows. The airspace must breathe as the air changes pressure due to daily fluctuations in the temperature and barometric pressure. If the airspace breathes warm humid air freely, there will be a buildup of condensation—with no countervailing reversal—until spring. The cure: seal the leaks of the inner pane and encourage leaks through all the others. Walls trap moisture in the same way and respond to the same treatment.

A prominent arctic expert once referred in public to the design of some dwelling huts to be used for a camp intended for very high latitudes. He was proud of the huts, for he had contributed to the design. "None of this scientific stuff for us," he said. "We will put the vapor barrier on the outside. We will wrap the huts in Visqueen to keep drafts out."

I pricked up my ears at that. Certain that moisture would pass through the walls and load them up with ice, I intermittently sought information from the users of the huts. Within a year or so, I found to my surprise that the huts worked fairly well.

"It's true that vapor passed through the insulation," one of the men remarked, "and it did condense and freeze on the plastic. But the plastic was fairly loose and every time the wind blew, the ice broke off into fine, little chips that fell between the plastic and the wall. Since we had to keep those huts pretty hot, the ice came in handy for making a Daiquiri now and then."

I was stupefied with admiration.

or leave. If the space "breathes" warm, humid inside air, condensation will occur.

It's about that simple: if the window space can breathe mostly outside air it will remain clear throughout the winter. If it breathes mostly inside air, it will stay frosted no matter what else is done. The solution is clear, then. Seal the potential leaks from the warm side, and encourage air passage from the cool side. For when a space breathes in, it should bring in cold (and therefore dry) air, which warms somewhat and becomes capable of carrying out with it any inadvertent moisture it may accumulate. If it can inhale warm, moist air, on the other hand, the air will cool. Excess moisture must remain between the panes, and the condensation buildup will continue as long as cold weather continues.

This breathing phenomenon is by no means confined to windows (in fact, it doesn't even occur in windows, if they are the factory sealed kind): it is the chief mechanism by which walls become wetted, saturated, and icy.

Truth and Consequences

The following story is often encountered in the northern winter: After a deep and soft January snowfall, in a school, apartment, restaurant, or office, a "leak" develops, staining the ceiling and perhaps causing a drip. A maintenance man is called, whose response is to provide buckets, after which he attacks the roof with a shovel. He removes the snow, which, sure enough, has turned to water at its lower layer. By the time he exposes and dries out the roof membrane, there's usually something to patch (Klutz's law: "If you shovel a roof, there'll be something to patch"). He patches it, and the leak stops. Everyone resumes normal activity, secure in the belief that watery snow leaked through a hole in the roof, and that the hole is now closed.

This happens to a degree in many places, all over the continent above about latitude 45, but it is endemic in the far North. It need not happen at all, of course. More likely than not, the roof did not leak (until the shoveler appeared). The drip was water trapped as frozen vapor in the insulation above the ceiling. The snowfall, by increasing the insulation, caused the melting of the trapped ice (a "January thaw" would have done the same). The melted ice, of course, dripped down as would water from any source. The reason the leak stopped after the roof was cleared of snow was that (1) the condensate was used up or (2) the remaining ice stayed frozen because the insulating snow was removed and the roof cooled off again. Even if there had been no patching of the roof, mere shoveling would have stopped the drip—until warmer weather occurred. For additional imprecations against the too-quick roof shoveler, see Seifert, "Attics and Roofs for Northern Residential Construction," HCM-00559.

Of course, there are other horror stories about what vapor can do in a cold climate. Fibrous insulation in a wall has been known to become saturated and slump to the bottom of the stud space, leaving the top few feet uninsulated. Vapor trapped under a membrane roof has caused blisters, which stretch the membrane and render it vulnerable to breaks and leakage. Winter-trapped vapor, escaping to the outside during summer, has caused paint to blister and peel, ice buildup has forced foam insulation boards from walls, and so on, on, and on. Vapor turns out to cause much more important problems than many temperate zone designers realize. For most such problems, the first thought of the user is to blame the contractor; the next thought is to blame the climate or bad materials. The fault, if blame need be assigned at all, is in the ignorance of the designer; vapor problems are not so hard to solve.

Why Don't We Have These Problems in Denver?

These same tendencies for vapor buildup occur also in lower latitudes. But they are less likely to cause noticeable problems there because the amount of water transferred is much less in these warmer areas. The average temperature difference is naturally smaller during the heating season in the Lower Forty-eight. Moreover, the heating season is short and is punctuated with warmer periods, during which condensed vapor can re-evaporate. The walls and ceilings never get very wet, and they have a chance to dry out fairly often. By contrast, in the Arctic, the heating season may approach 12 months in length, and the drying-out of insulated spaces may never take place. In cold regions, therefore, moisture buildup in poorly designed structures dominates during winter. During the relatively short summer, the insulation can only partially dry out. Farther north in the Arctic, moisture buildup can be a one-way proposition, rarely, if ever, improving. And yet, the same design may be quite tolerable in warmer climates.

The Cure

To avoid moisture problems, it is fairly obvious that improvement, if not cure, can come from the following:
1. enough ventilation to keep the relative humidity low on the warm side of the wall.
2. a leak-free vapor barrier on the warm side, and
3. highly permeable materials on the cold side.

The first of these, ventilation, must be used to a certain degree, if only to keep the humidity at an acceptable level and the air "fresh"; too-dry air causes some people bronchial discomfort and possibly health problems.

There are other reasons, too, for curtailing unlimited ventilation. For one thing, ingesting vast quantities of outside air merely for the purpose of exhausting it to carry a little water away is enormously costly and wasteful of fuel. (It's possible to mitigate this cost somewhat by using elaborate heat exchangers—but these ain't so cheap and trouble-free either.) For another, keeping the air too dry does awful things to furniture: heirloom antiques crack, pianos become disjointed, chairs and tables shrink, and their wood tends to check. Even after a satisfactory adjustment has been made to all this, later, during the short summer, "reasonable" humidities prevail and wooden parts re-expand, to repeat the drying cycle yet again the following winter, ad infuriatum. So it is not likely that we will elect to keep the air totally dry, even if that were possible. In any case, the warmed air will pick up quite a bit of water, and some of that will certainly migrate to the interior of walls and roofs, barring some additional effective measures.

The second part of the cure, a leak-free vapor barrier, is absolutely essential. Without one, the other forms of vapor control cannot be adequate. There is no such thing as a permanent arctic enclosure without a good vapor barrier on the warm side. Trouble is, no vapor barrier is totally tight.[3] There are leaks despite whatever precautions we take. Sometimes the leaks are tremendous, as happens where a barrier is discontinued near a door or window opening, or where it may be lapped poorly where wall and ceiling join. Often the vapor seal has openings in it where electrical outlets or fixtures occur. Various types of pans are available for sealing the electrical outlets, plumbing, or other penetrations to the vapor barrier. The thing is to provide as tight a vapor barrier as can be installed conveniently and to recognize that it will be imperfect.

To take care of the inevitable leaks, it is necessary to provide for high permeability of all layers of construction colder than the vapor barrier. Ordinary construction is usually not porous enough and must be perforated in some manner so that "breathing" takes place freely to the cold side and not at all to the warm side. Usually, the equivalent of five or six half-inch diameter holes per stud space is enough, recognizing of course that such openings should be protected from entry of driving rain, snow, or insects. Porous sheathing is available and is fine.

3 One public construction agency, well aware of the enormous expense entailed in tolerating poorly designed walls, has specified metal foil laminated with kraft paper as the vapor barrier. It is to be sealed at all joints and inspected to ensure against leaks whenever possible. This method works very well indeed. It is my opinion, however, that the cost is far too great, that it is possible to make equally passable walls with much less expense.

Figure 4. Placement of vapor barrier. Vapor barriers should be on the warmest practicable surface. The vapor barrier at (1), however, has to be penetrated by electrical wiring and fixtures and so is usually leaky. In some cases, arranging for the barrier to be at position (2) has merit. The barrier is safer there, but it could cool below the dewpoint for short periods. Another advantage is the reduction of all-wood heat flow paths.

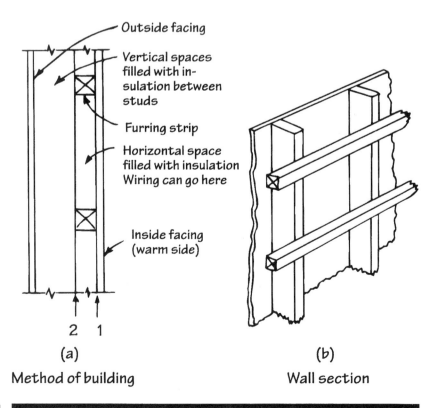

Outside facing
Vertical spaces filled with insulation between studs
Furring strip
Horizontal space filled with insulation
Wiring can go here
Inside facing (warm side)

2   1

(a)
Method of building

(b)
Wall section

Vapor Barriers and Insulation

Because vapor problems do occur even in temperate environments, manufacturers of insulation frequently make "blankets" or "rolls" of insulation with a vapor barrier of tarred paper and/or metal foil attached. Such vapor barriers are admirably impermeable, and if diffusion were the only mechanism for vapor transport, they would be superb. Even as they are, they perform satisfactorily for temperate zone conditions. But for cold climates this "foil-backed" insulation is not adequate because the numerous joints used in construction permit surprisingly large air leaks. Furthermore, the foil, which has some value as a thermal barrier when facing space, has none when in contact to any building material and there is little virtue in leaving an air space unfilled with insulation in the cold regions. For an adequate and cheap vapor barrier, large sheets of 4 to 6 mil or thicker polyethylene sheeting have proven to be satisfactory. Such sheets can be generously lapped and taped where they join and have the further merit of being almost transparent, so that the condition of the insulation beyond can be checked after installing them on the inside walls.

A large-sheet vapor barrier not only discourages leakiness, it also allows the abandonment of foil-backed insulation, which is costly and troublesome to install. Instead, "friction-fit" batts may be used, and they can be installed so quickly that the entire vapor-barrier-plus-insulation package for an adequately designed wall will usually turn out to be cheaper than the alternative, inadequate, foil-backed system. The choice should be obvious, but many builders and building material dealers continue to use or to recommend foil-backed insulation for arctic and subarctic work. Possibly they believe that "if it costs more, it must be better." If so, in this case they are mistaken: a well-designed wall with a large sheet plastic vapor barrier over friction-fit batts filling the stud spaces is in every way superior to one with only foil-backed insulation as a vapor barrier, and the fact it is somewhat cheaper and quicker to install should not be held against it.

A few simple design precautions yield admirable results:

1. Use large vapor-proof sheets near the warm side of an outer wall or roof insulation.
2. Seal the holes and joints in it.
3. Assure that materials colder than the vapor barrier are able to breathe to the outside air only.
4. Calculate to make certain that the desired relative humidity is possible for the design. If it is not, adjust either the design or the humidity.
5. Where walls are of masonry, consider placing insulation on the outside.
6. Consider an "upside down" roof, with the waterproof membrane serving also as the vapor barrier (Figure 5).
7. Avoid any possibilities for multiple vapor barriers (these form vapor traps).
8. Operate the building at a neutral pressure so as to not force warm moist air out through cracks in the wall and roof.

Figure 5. Roof systems. Most flat roofs are built as in (a). This design is satisfactory when the underside of the roof membrane is well ventilated, and when the ventilation is augmented by a warm-side barrier. Section (b) represents the IRMA roof of the Dow Chemical Company. There is no vapor trap. It is limited to waterproof insulation.

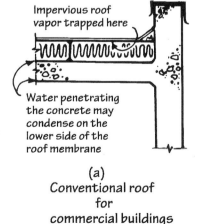

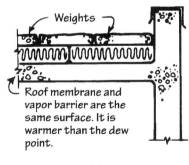

(a)
Conventional roof
for
commercial buildings

(b)
Upside-down roof
for
commercial building

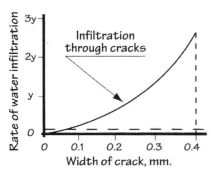

Figure 6. Relative effect—Diffusion as compared with convection through cracks. Even relatively small leaks admit moisture at magnificent rates. For vapor barrier materials, the leaks become a much more effective avenue of vapor flow than does the diffusion through the body of the material. Diffusion is not the dominant mechanism, especially for relatively good vapor barrier materials. Leaks are more important.

Conclusion

Vapor transport into walls and roofs creates more trouble in cold regions than elsewhere, but once the conditions of its accumulation are understood, the problems can be handled quite adequately, without even recourse to excessive costs that usually accompany polar work.

It's a pity, but if the designer of Arctic buildings does his homework, understands the fundamentals of vapor transport, and designs properly to avoid problems, his contributions will be entirely unnoticed by the casual user.[4] But capable engineers have always worked under that handicap: successes are architectural or scientific triumphs, while failures are always "engineering" failures. To my way of thinking, however, the trouble-free installation is the mark of the knowledgeable engineer with a command "of the basics." There is no greater hero than he. For a good designer to remain unnoticed is often the greatest of compliments.

References

ASHRAE. *Handbook of Fundamentals.* American Society of Heating, Refrigerating and Air Conditioning Engineers, Atlanta GA. 1972.

Seifert, R. D. *Attics and Roofs for Northern Residential Construction.* Alaska Cooperative Extension Publication HCM-00559.

Croome, D. J., and A. F. C. Sherratt. *Condensation in Buildings,* London: Elsevier Science, 1972.

Hutcheon, N. B., and G. O. P. Handegord. *Building Science for a Cold Climate.* Construction Technology Centre Atlantic, Fredericton, New Brunswick, Canada, 1990.

Hutcheon, N. B. Humidified Buildings. *Canadian Building Digest 42.* Ottawa: National Research Council, Division of Buildings, June 1963.

Hutcheon, N. B. Humidity in Canadian Buildings. *Canadian Building Digest 1.* Ottawa: National Research Council, Division of Buildings, January 1960.

Latta, J. K., and R. K. Beach. Vapor Diffusion and Condensation. *Canadian Building Digest 57.* Ottawa: National Research Council, Division of Buildings, September 1964.

McFadden, T. Moisture Accumulation in Insulation Systems of Buildings and Residences. ASCE *Journal of Cold Regions Engineering 3,* September 1988.

McFadden, T., and F. L. Bennett. *Construction in Cold Regions.* John Wiley and Sons 1991.

Miller, W. T., and F. B. Morse. *Excessive Heat Moisture in Homes.* Purdue University: Engineering Extension Service, 82, 1953.

Rogers, T. S. *Thermal Design of Buildings.* New York: Wiley & Sons, 1964.

Seiffert, Karl. *Damp Diffusion in Buildings.* New York: Elsevier, 1970.

Wilson, A. G. Condensation on Inside Window Surfaces. *Canadian Building Digest 4.* Ottawa: National Research Council, Division of Buildings, April 1960.

Wilson, A. G. Condensation Between Panes of Double Windows. *Canadian Building Digest 5.* Ottawa: National Research Council, Division of Buildings, May 1960.

---

4 Maybe it would be a good idea to ensure that at least one enclosure on each major installation be designed by a designer inexperienced in cold-climate problems. The contrast may provide the arctic builder's only chance for recognition.

Northern Construction: Windows

CHAPTER 6

Introduction

Windows have so many virtues that architects and engineers rarely contrive enclosures without them. Windows primarily are to provide a view—that's the one consideration that many people consider necessary and sufficient. But they also admit light (and air, if they can be opened). They can be arranged to trap solar heat and, of course, they serve as useful emergency exits in case of entrapment. Windows are altogether desirable and every architect prides himself on his skill at fenestration.

And well he may, for designing windows properly (and choosing their number and size) is no routine task. Nor is the task always done well, for there are special vices that windows possess that mitigate their virtues. First of all, windows are far more expensive than is generally supposed. Windows at best cost quite a bit more per square foot than walls. Where heat flow control is important, as in heated (or cooled) structures, the annual cost for windows may be from ten to twenty times as great as for well-insulated walls of equal area, even when washing and replacement costs are not considered. In addition, the chief advantages of windows (view and light) do not even exist at night—that is, during most of the hours of arctic winter—yet the incredibly high heat loss costs continue at night just as if windows were fulfilling their purposes. Moreover, windows have cool surfaces in winter, a circumstance that not only causes discomfort to occupants but also causes the collection of dew or frost with a whole host of accompanying problems.

For these reasons and others, selecting windows provides a promising area for refinement in the interest of utility, fuel savings, comfort, and cost. Designers must be familiar with facts about windows lest they inadvertently sentence their clients to high and irremediable annual costs; and these costs, far from representing value paid for, may actually have bought additional discomfort and trouble.

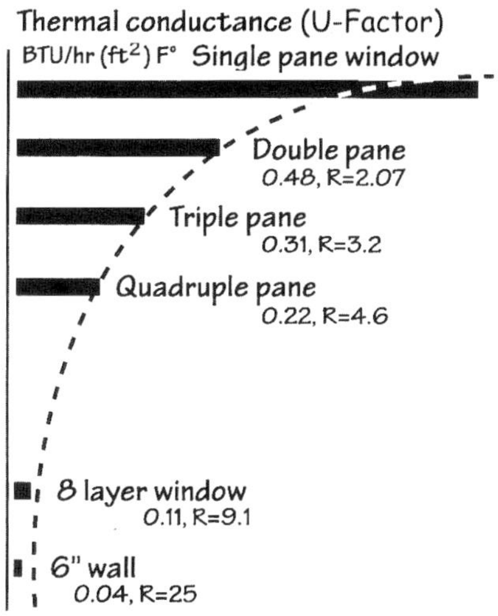

Windows as Heat Barriers: 1, 2, 3, 4 ...

Windows are a lot better at resisting heat flow than an open hole in the wall, but they are a lot worse than a properly designed wall. A window composed of a single layer of glass not only stops air exchange through the opening it fills, it also forms a barrier to heat conduction. Conductive heat flow is resisted by the glass or plastic itself to some extent, but mostly by the two films of air that "wet" the surface of the pane. In the case of two or more parallel panes, the air spaces augment the films as the primary sources of thermal resistance. The basic heat flow relation is that the flow, given in BTUs per square foot per hour (or in watts per square meter), is inversely proportional to the sum of the

Thermal conductance (U-Factor)
BTU/hr (ft²) F° Single pane window

Double pane
0.48, R=2.07

Triple pane
0.31, R=3.2

Quadruple pane
0.22, R=4.6

8 layer window
0.11, R=9.1

6" wall
0.04, R=25

Figure 1. Windows as thermal barriers. It is impractical to expect to combine transparency and good thermal resistance. Even with eight successive sheets of transparent material, spaced three-quarter inch or more, a window conducts nearly three times as much heat as an insulated, six-inch wall.

conductive resistances of the material of the window, the air spaces, and the air films:

$$\frac{q}{A} = \frac{1}{\sum R}(\Delta T) = U(\Delta T)$$

"R" may be closely estimated from values reported in various reference works. The quality of a window to resist heat flow is typically stated as the U-Factor, $U = 1/\sum R$, and varies with the wind, the mean temperature, the temperature difference, the number of successive panes through which the heat must flow, and the size and shape of the window. The lower the value of U, the better the window at resisting heat loss. Typical values are shown in Figure 1. The values are figured for a mean temperature of 0°F, a temperature drop across the air spaces of about 20°F, and a 15 mile-per-hour wind: about right for arctic conditions.

The thermal resistance of an air space is influenced by the shape and size of the space, for there is convective and radiative heat transfer in the problem masquerading as conduction (Figure 2). R is also influenced by the reflectivity and emissivity of the materials bordering the space. Very thin spaces have less resistance value than thick ones, for in thin (3/8 inch or less) spaces conduction allows heat to be transferred readily. Of course, both convection and conduction are always operative, as well as heat transfer due to radiation.[1] In airspaces thicker than about 3/4 inch, convection heat transfer begins to dominate, and little thermal resistance is gained by increasing the width of the airspace.

1 Radiation heat transfer accounts for about one-half the heat loss through two-pane windows when the spacing is greater than one-half inch.

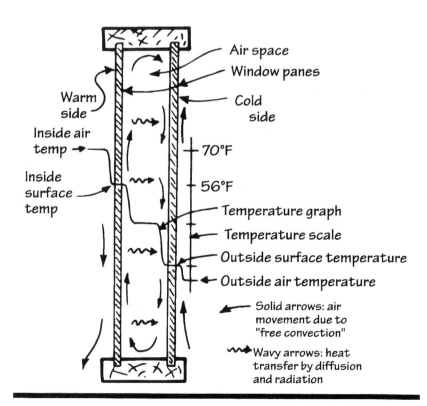

Figure 2. Heat flow through multiple pane windows. There is circulation of air in the spaces between panes. The warmer pane (here shown at 56°F) is colder than the 70°F room air but warmer than the 42°F between panes. Similar temperature differences across the colder pane induce "free convection" currents. In thin spaces (less than ½ inch), conduction dominates and accounts for a disproportionately high heat loss. Thicker spaces are better.

It is well to observe, too, that skylights are considerably worse than vertical windows.

In a manner analogous to the selection of the optimum wall thickness, one can determine the "best" number of panes to use. Roughly, double panes pass only half as much heat as would flow through a single sheet of equal area. It would take four panes to halve the heat flow yet again, and so on (see Figure 1, page 56). A simple calculation will reveal the "best" configuration: compare (1) the value of the heat saved each year by an additional pane with (2) its annual capital cost. Continue to add panes until the costs exceed the benefits and select the configuration yielding the lowest total annual cost. In most arctic areas, using present costs for fuel oil, labor, and materials, windows generally prove to be most economical when composed of three or four multiple panes. Fewer panes increase annual cost for heat; more panes increase annual cost of the investment. Remember that even after the optimum number of panes is selected, the total cost (though a *minimum* for windows) is *much* greater than for optimum walls.

How about Using Solar Heat?

It is one of the great ironies of our time that one of the disadvantages of large windows in the Arctic is heat gain, not heat loss. (As I write, in early April, the temperature in my office is over 80°, with my windows and door wide open for relief from heat-stroke.) For at high latitudes, even the summer sun, at its highest, may be only 30 or 40 degrees above the horizon. This means two things: (1) the sun hits vertical windows "head-on," and (2) mechanisms for shading glass so often used in warmer climates (horizontal overhangs, "cool-shade" screens, etc.) do not work as well as they do where there is a higher sun altitude. The result is that the glare from large expanses of glass may be intolerable and the heat gain insufferable. It is a common sight to see large and incredibly expensive areas of glass lined with aluminum foil in an effort to reduce heat gain and glare to tolerable levels.

Naturally, heat loss goes on just as if the windows were transparent. Heat gain, of course, can be reduced by using metallic reflectors: anything that will reflect light back through the windows at visible wavelengths (to which glass, naturally, is transparent) will reduce heat gain. That requires a metallic, or specular, reflector. Otherwise, the "greenhouse effect" controls: light, passing through transparent windows, strikes and warms surfaces within. These must reradiate the incoming energy, but at the longer wavelengths appropriate to their increasing temperature. Windows are not necessarily transparent to those (infrared) wavelengths, nor are walls, ceilings, floors, and furniture. At the ultimate, the heat gain can be enormous, as anyone will attest who has experienced a tightly closed automobile on a sunny warm day. Careful designers can make good use of this free heat in the North, but it is rarely done well because of the difficulty of control. And, of course, the heat is only available for limited periods on bright days.

So there you have it: windows don't work at all during the dark of winter and they work entirely too well during the long, bright days of summer. Moral: when you buy windows, it's wise to know what you're paying for and what you're getting.

How about One-way Mirror and Low-E Windows?

Manufacturers know all these things and more, so they have offered "heat-reflecting glass." This works well: the glass is made like a half-silvered mirror (usually the metallic coating is on one or more of the inside surfaces of a factory-sealed double-pane unit, for mechanical protection). Most of the incident visible light (sunlight) is reflected, yet there is no problem seeing out. In the daytime you can see out but not in; and in the night, in but not out. Such glass mitigates the twin problems of heat gain and glare very well indeed. Unhappily, converting windows to one-way mirrors does nothing at all for heat loss nor for condensation, and these are the two controlling aspects in window selection and design.

A more recent window design incorporates a low-emissivity (low-E) coating that reduces the radiation transfer between the panes. Argon or krypton gas replaces dry nitrogen between the panes. These techniques yield R-values of 3 for two-pane low-E windows to 4 for two-pane low-E and argon-filled windows.

Condensation

Quite likely the most notable thing about the winter window is its tendency to fog over. Moisture precipitates as dew or frost on a cold window pane. Later, perhaps, the moisture runs off to form wet areas on window sills and under window walls, which can lead to an expensive maintenance problem and a great bother. An even greater nuisance is that a window coated with frost or dew is no longer useful for seeing through. Sometimes people find themselves continuously engaged in scraping the frost from windows to maintain a sanity-preserving view, with occasional time off to tend a humidifier.

Windows as Dehumidifiers

A window, as the coldest available interior surface, must collect moisture until the warm air sweeping across it reaches a dryness appropriate to the saturation limit for the window's inside surface temperature, because the vapor content of a space is controlled by the coldest surface within it. For example, we may imagine a typical double-pane window when the outdoor air temperature is −30°F. If the indoor temperature is 70°F, what is the greatest relative humidity permissible if fogging of the window is to be avoided?[2] A rough calculation reveals that the temperature of the inside (foggy?) surface of the window is somewhere around 45°F. Air saturated at that temperature would be found from a psychrometric chart to have a relative humidity of about 40%. Room humidity greater than 40% would therefore be the practical maximum. Should the moisture in the room exceed this, it would continue to condense on the window, which would continue weeping until the relative humidity in the room is restored to 40%. Had the window been a single pane, the window surface would have been near +10°F. In that case, the precipitation would be in the form of ice, and the room would have to be kept at

---

2 Refer to the preceding "Northern Construction: Vapor Barriers," p. 45 for a treatment of moisture migration.

10% relative humidity or lower to avoid it. With three panes, the inside surface temperature would be around 52°F, and the relative humidity could be as much as 50% before fogging occurs.

It is possible to maintain higher relative humidities, if one is willing to undertake heroic measures. Pet shops, flower stores, swimming pools, baths, and laundries are all examples where function or comfort requires high relative humidities. To maintain clear windows in such places, one may raise the inside window surface temperatures by blowing room air

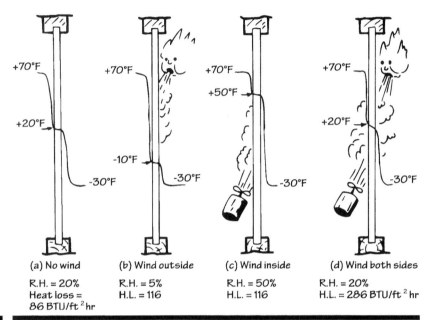

Figure 3. Wind increases the heat flow rate dramatically and influences the permissible relative humidity. The influence is favorable (no fog) if on the inside, unfavorable outside. With wind on both sides, the heat flow is more than tripled. The effect is less for multiple-pane windows.

(a) No wind
R.H. = 20%
Heat loss =
86 BTU/ft² hr

(b) Wind outside
R.H. = 5%
H.L. = 116

(c) Wind inside
R.H. = 50%
H.L. = 116

(d) Wind both sides
R.H. = 20%
H.L. = 286 BTU/ft² hr

across the surface. This is costly in terms of energy loss, but it helps reduce fogging. Figure 3 illustrates the point. Had the Figure 3 example been for double glass, the fan's effect would have been equally dramatic.

The permissible relative humidity would have been around 40% with no fan, and as high as 75% with a fan washing room air over the inside surface. Obviously, at some cost in heat, a fan can do wonders for steamy windows, and those wonders come easier when the window is multiple-paned.[3]

In chapter five, it was shown how vapor could accumulate between sheets of a multiple pane window. The "cure" was to ventilate such spaces to the outside while keeping the inside pane sealed. A more common way to avoid such inaccessible condensation is to use sealed units that are permanently filled with dry gas during manufacture. Such units are relatively trouble free. In the Arctic it may often pay to use the high-performance low-E argon-filled unit, making sure that this is the economical solution. (Factory-sealed double and triple glazing units usually have air spaces around $1/2$ inch. There is a penalty of about 5% for reducing air spaces in multiple glazing from $3/4$ to $1/2$ inch, whether or not the spaces are hermetically sealed.)

3 In automobiles, where clear vision is important, strong "defrost" fans are used in just this way. If the windows persist in fogging, opening a window or door long enough to replace the warm humid air will usually clear things up.

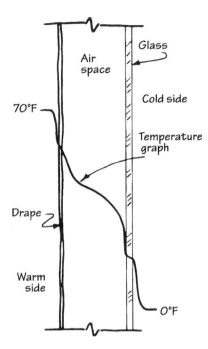

Figure 4. Effect of drapes and hangings. Heat flow, with drapes present, is about 32 BTU/hr ft². Heat flow, drapes absent, is 82 BTU/hr ft². But the surface temperature of the window has dropped from 53°F to 14°F. Result: icy windows in the morning, which will melt and run down on the windowsills after the drapes are opened.

Vice Mitigated: The Arctic Builder Can Beat the System

The winter window, it is apparent, (1) doesn't do much for light or vision at night, (2) irresponsibly delivers heat to the outside, (3) collects moisture, and (4) costs a bundle. Yet in the daytime, it can admit light and even air (if it isn't frosted shut). And it serves as an essential emergency exit or auxiliary air source. Can the vices be mitigated without sacrifice of the virtues?

Consider this: at night, a window is no window—it is merely a fragile, expensive, damp-collecting wall. Would it not be desirable to place insulation over all or part of a window at night, to be opened during the day, in much the same manner as wooden shutters were used in former times? Yes, of course. For a price, insulated panels can be provided to cover windows for the times when windows are unwanted, to be removed when sun, light, or a view are desired.

The catch: To an extent, drapes and blinds serve such a purpose. They reduce glare and heat loss (but do nothing for heat gain) and they are readily drawn aside when windows are wanted. Could they not be made thicker, say, as are winter garments?

Of course, but to do so lowers the window temperature, and when the drapes are drawn open, they reveal only an expanse of ice where a view was expected (Figure 4).

The solution, sad to say, is that the insulation must be on the outside to avoid frost buildup. That will require strong, rigid shutters that can be opened and closed easily (or removed and stored easily) and that can withstand winter winds. It's a solution, all right, but one that will require a fair share of attention to structure as well as to esthetics. Never-

Figure 5. Shutters are not normally hermetically sealed as a window is. So the window glass is allowed to cool below the dew point (as it must be if it is on the cool side of a heat barrier), and condensation will occur. Shutters or storm windows should therefore be on the outside only.

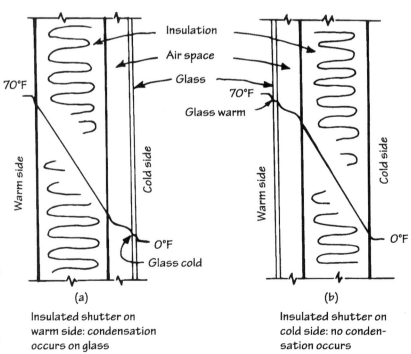

(a)
Insulated shutter on warm side: condensation occurs on glass

(b)
Insulated shutter on cold side: no condensation occurs

theless, it is a worthwhile thing to strive for, in the hope that structural and thermal adequacy will also be affordable (Figure 5).

Conclusion

There is every reason to be cautious in the design and selection of windows for Habitat Arctic. The aim is to ensure that windows do all that windows are supposed to do and to minimize the expensive side effects that too often accompany poorly designed ones. Heat loss can be reduced by using low-E multiple panes, and this serendipitously helps cure moisture buildup as well. But even at its best, the transparent window is a failure as a wall. Perhaps at night, or in storms, or when occupants desire enclosure, when a window is not a window, some form of removable insulated covers (shutters) could be used. If such were provided, single-pane or double-pane windows quite possibly would again be the economical solution and the overall energy loss and comfort problems would be minimized, too.

Do not overlook the frame. Most windows destined for commercial buildings use extruded aluminum frames. All the comments made with respect to condensation on the glazing applies to the frames as well. Be certain that the metal frames have an adequate thermal break to avoid condensation on the frames.

Windows are too important to be casually specified. They should be designed with the consequences firmly in mind. For some applications, windowless structures may be acceptable (large stores, swimming pools, and theaters, to name a few), but for living and working quarters, windows seem to be a psychological imperative. Let us be certain they do their work right.

References

ASHRAE. *Handbook of Fundamentals.* American Society of Heating, Refrigeration and Air Conditioning Engineers, Atlanta, GA.

Croome, Derek J., and A. F. C. Sherratt, *Condensation in Buildings.* London: Applied Science Publishers, 1972.

McFadden, T. and F.L. Bennett. *Construction in Cold Regions.* New York: John Wiley and Sons, Inc. 1991.

Rogers, Tyler Stewart. *Thermal Design of Buildings.* Wiley, 1964.

Seiffert, Karl. *Damp Diffusion and Buildings.* London: Elsevier, 1970.

Water Supply in the North

CHAPTER 7

Introduction

Historically, finding water in the polar regions has been no problem: nobody dies of thirst in the Arctic, and cartoonists have never felt impelled to exercise their pens in describing mirages or hallucinations seen by the thirst-crazed polar traveler. Yet the total annual precipitation of water in the high Arctic is slight—about the same as that of the deserts in the Southwest United States. Travelers have been able to survive in the Arctic without canteens strapped to their belts and with no other liquid containers than those used for cooking because in the summer, poor surface drainage ensures an abundance of water in most areas; in winter, snow and ice guarantee a source of water whenever there is enough energy to melt it.

Some Native American groups in early times considered it a matter of hospitality and prestige to be able to offer guests water or soup. In later years tea or coffee at mealtimes was considered essential for the Arctic traveler. This was virtually all the liquid consumed, for a person on a winter trail is not eager to carry more water than one can drink.

But with the advent of permanent dwellings, fabrics that need laundering, utensils and dishes that need to be washed, and with changing ideas of personal hygiene, water has become a problem: there are, in the Arctic, problems with water quantity and quality and its conveyance and storage. While it is unlikely that water consumption in the Arctic will ever need to match that of warmer regions (fewer lawns or fountains), still, provisions for water supplies for dwellings, communities, or camps require a mastery of all the ingenious techniques used in temperate regions plus the ability to solve problems unique to cold climates.

For example, groundwater, as found in subarctic and warmer zones, is not to be expected where the permafrost is deep and continuous. Nor is surface water easily come by. Away from large lakes and rivers, surface water may be available in summer only. As for rain barrels, catchment of liquid water will hardly ever be worth the trouble. The quarrying and melting of snow or ice is a possibility, but becomes practical only when there is heat to spare (e.g., as waste heat from engines or power plants) or when there is no other alternative.[1]

Nevertheless, the problems are solvable. Usually, the expense is great—and is made even greater by the difficulty of satisfactory disposal of wastewater—so there should be the utmost effort to avoid wasteful uses of water in the human environment. To name the obvious case, flush toilets of the conventional kind are to be deplored: In many arctic installations, toilet flushing uses more than half of the total supply of drinking water. In an era where water-conserving

1 Outdoors men (outdoorspersons) are sometimes advised to melt their own drinking water by carrying ice in a waterproof bag inside their clothing. This heroic technique is used only when there is no reasonable alternative. It beats dehydration at the expense of potential hypothermia.

commodes are readily available, the selection of the ordinary flush toilet is never justified.[2]

Recent experience in construction camps in the Arctic has shown that water consumption ranges from 10 to 100 gallons per person per day, and these figures of course do not include normal city uses like street and vehicle washing, lawn watering, fountains, or industrial consumption. Until such uses are extended to the Arctic, it's probably safe to provide a water supply of 80 gallons per person per day plus whatever more is needed for fire protection.

Groundwater

Groundwater is not to be found by the dowser with a fork of witch hazel. Where continuous, permafrost tends to be deep and the water under it is often saline or of poor quality. The high cost of drilling for water, conducting it through several hundred feet of frozen ground, and then treating it for potability discourages the search for groundwater under the permafrost except in special circumstances. Figure 1 is a generalized block diagram of permafrost terrain, showing that perched aquifers may be expected to underlie small lakes and rivers and that under large lakes there may be a thawed water-bearing substrate all the way through the permafrost. (Of course, where there are large lakes, they themselves become suitable water supply sources and their potential as groundwater indicators is not important.) Thawed aquifers under rivers have been exploited by comparatively shallow wells to provide a year-round source of water of good quality. Some people may quibble over whether such water is true groundwater or merely surface water gathered through an infiltration system: the point is unimportant—such sources may be the only continuous supplies available in the true Arctic.

At lower latitudes, where the permafrost is thinner and possibly discontinuous, subpermafrost water is likely to be found; and in areas away from the seacoast, it may be a satisfactory source—or the only practicable source. In cases of subpermafrost supplies, there are of course some serious problems to solve in keeping the well shaft unfrozen. And when it is kept unfrozen, there may be problems of subsidence and stress due to inadvertent melting of ice-rich ground around the well pipe.

2 See *The Northern Engineer* summer 1973 issue devoted to this problem, especially Amos Alter's article, "The Polar Palace," and Gina Brown's Editorial.

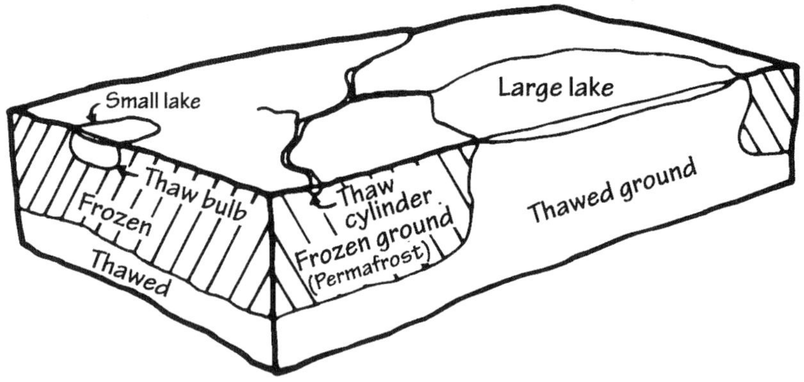

Figure 1. Large, deep lakes may have bottoms that are thawed all the way through the permafrost. Smaller lakes and streams have thaw bulbs or thaw cylinders perched within the frozen ground. Thawed areas may serve as aquifers.

Surface Water

Surface water supplies for major uses in the high Arctic are limited to two types of sources: lakes and streams that do not freeze to the bottom and streams that, even when frozen to the bottom, have an adequate year-round flow below the stream bed. To be useful for winter water, should have plenty of water beneath the ice. There are a great many lakes in the Arctic; however, they are mostly much too shallow to serve as dependable all-year reservoirs, even when summer recharge can be assured. Efforts have been made to deepen lakes by raising their outlets and to trap additional water on their catchment areas by erecting snow fences. This works. It may even be the economically desirable solution to augment water supply where liquid water is scarce.

Streams provide a source of good-quality water in the summer, and if large enough to retain a thawed cylinder of gravel, may support a continuous flow all winter long as well. Most rivers in the Arctic tend to have braided channels and shallow water, and it can be difficult to predict where an unfrozen channel will be and whether it will remain open to the main under-ice flow aquifer. Channels are often pinched when frost deepens during winter, and thus much of the surface ice is formed when water, forced upward through cracks, builds up and freezes to the top of the ice already formed. Such aufeis ("glaciering," "icing") may form over surprisingly large areas (some individual icings cover as much as three or four square miles) and may be up to 25 feet thick. The melting of such deposits is important in sustaining summer stream flow. Since such ice is usually frozen solidly to the bottom, meltwater tends to run over the top of the ice, rather than to lift and break ice in the tradition of the "spring breakup" of larger or more southerly rivers.

Winter flow of streams whose catchment area and course lie over continuous permafrost is only a tiny fraction of the summer flow, which in turn is small compared to the spring peak (Figure 2). Small streams in such country are totally immobilized in the winter as soon

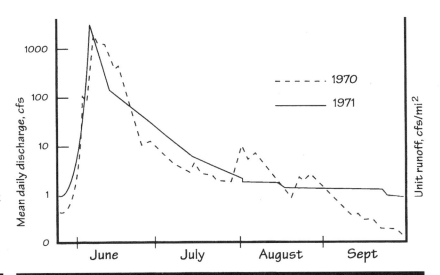

Figure 2. Runoff from the Putuligayuk River, a typical stream in the Arctic Coastal Plain. Note that the peak runoff is from meltwater in June, and that by July, runoff is less than a thousandth of this.

as the active layer is solidly frozen to the permafrost. We do not fully know by what mechanism, or from what reservoirs, winter flow manages to reach the thaw bulbs of the rivers, but it is certain that some of the flow emerges from thermal springs in the headwaters. This, plus drainage from the lakes whose thaw bulbs intersect the rivers, probably accounts for most of the winter flow. Summer flow, of course, is from melting snow and ice or from rainfall upon the largely impermeable terrain. It is not yet known what portion of the water reaching large lakes penetrates the ground to recharge the subpermafrost aquifers.

What about Ice Itself?

Using ice to supply domestic water is one possibility where waste heat is available, as from electric generating plants. It is particularly expensive since ice must be quarried and transporting ice chunks is troublesome. If consumption is assumed to be around 50 gallons per capita daily (about half of the temperate zone average), over 400 pounds of ice must be quarried, handled, and melted for each resident each day. Such a heroic task is too much for the ordinary situation: Wherever possible, water is kept liquid so that it can be handled by simple pipes and pumps, and so that it can be used at the mere twist of a valve or pump control.

The heat to melt and warm 100 gallons of water is equivalent to the combustion of about 1.1 gallons of fuel oil. In places where fresh water is available only by melting ice or snow (as at the South Pole), the labor and fuel costs become so important that there is a real incentive to limit water use.

Nevertheless, melting ice for drinking and cooking is a practice of long tradition among dwellers of the coastal Arctic. For modest supplies of high-quality water, Eskimos have customarily collected blocks of sea ice from the upper parts of old pressure ridges. Such ice, having been through one or more summer seasons, is nearly free of salt and has no other significant contaminants. Commonly, it is stored as ice until needed, then someone brings a chunk or two inside to melt by house heat.

Peyton, Johnson, and Behlke (September 1967) have experimented with the natural desalination of seawater ice by spraying seawater into the winter air and allowing ice to build up in high mounds that allow drainage of the brine rejected in the freezing process. Their successful experiment points to a possibility of growing freshwater ice mounds out of sea water, with the idea that the summer melt from such mounds may supply good water in areas of short supply.

While technically feasible, however, the melting of ice for winter water is so costly and troublesome that the normal way of gathering water in quantity is to pump it from beneath the ice of rivers or lakes and store it in liquid form until needed. At Wainwright, at the British Petroleum Camp near Prudhoe Bay, and at other places, the winter's water supply is pumped from a lake in summer and stored in insulated, slightly heated tanks.

How Much Water is Enough?

When water is plentiful and good and conveniently available from pressure taps, water use merely for domestic purposes approaches 100 gallons per capita-day (GPCD). Experience in arctic construction camps reveals that, at a drilling site operating in March and April of 1971, water use ranged from 10 to 60 GPCD, with an average of about 35 GPCD. At a base camp (a camp with many amenities, including plumbing) in the same area, from October 1971 and January 1972, the range was from 35 GPCD to 90, averaging 53 GPCD (Tilsworth and Damron 1973). These figures are probably typical: the difference in consumption is due to the easier availability of water at the base camp. Oddly enough, the highest peaks of per capita consumption occurred in the coldest months, possibly due to extended hot showers by chilled people.[3]

In strong contrast, where water has to be carried by individual householders and there is no pressure plumbing system, water consumption ranges from about 3 GPCD to perhaps 10 GPCD. Figures much lower than these may be attained reflecting the difficulty in backpacking water as well as tolerance for long intervals between baths.

How Good is Arctic Water?

Naturally enough, the quality of water is highly variable from place to place. For example, water from lakes and ponds is often quite acceptable during summer, autumn, and early winter. But as the ice thickens, the dissolved and suspended impurities are refined out of the ice and the unfrozen water becomes thicker with the increasing concentration of salts and organic contaminants. The shallower the lake, the higher the concentration of unpleasant chemicals in the water, until in late winter and spring the natural supply (if it holds out) can become unbearable without some form of treatment.

Using groundwater has its difficulties too, but its quality varies less with the seasons. A common problem is salinity; another is iron. If water contains a tastable quantity of salt, it is hardly usable for domestic purposes. Yet on Alaska's Arctic coast, there are at least two installations where fresh, domestic water has to be distilled from the brackish water of near-shore lagoons. This is horribly expensive in money and energy, and the plants will surely be phased out as better sources become available. The individual householder could not reasonably expect to use brackish water because desalination costs are generally far too high when using very small desalination plants.

Iron seems to be a greater problem, however, to the uninitiated: Iron-bearing water plugs pipes and boilers with scale, causes ugly brown stains on plumbing fixtures, turns white laundry amber, turns whiskey violet, and turns house holders livid. As we will see, though, the circumstance is not totally hopeless.

Hardness is also common in groundwaters of the subarctic, and communities like Fairbanks have chosen to soften the entire municipal supply. And of course, any water used in steam boilers will likely require extensive softening or demineralization prior to use. Glacial streams carry large quantities of rock flour and many of the larger

3 Some of the peak flows might have been augmented by the wastage of water to prevent freezeups. Periods of low consumption may reflect occasions when even such wastage was unsuccessful, and the pipes froze.

rivers carry this, as well as much silt and organic debris. A fairly high pH is common in these waters, and this, together with low temperatures, makes even simple chlorination not so simple. In contrast, the tea-colored drainage from muskeg and swamp may be super-soft and somewhat acidic. Such water is aggressive in its corrosive effect on iron plumbing and utensils (Clark and Alter 1956).

Water obtained from the melting of specially quarried ice (from sea-ice ridges or from fresh water sources) is potentially the purest water available. Unfortunately, the process of finding, cutting, transporting, storing and finally melting the ice introduces so many opportunities for contamination that real (as opposed to ideal) supplies of water from ice are often of questionable quality.[4] There have been several outbreaks of gastrointestinal illness reported from the use of contaminated ice.

On the other hand, travelers in or on the polar pack ice in late summer and autumn find ponds of pure water covering perhaps 20% of the total area of thick sea ice. Such water may rank among the world's purest, if it has not been visited by a transient walrus or polar bear. Unfortunately, it is available only to the relatively few odd characters who travel on or through the sea ice.

Water in the North is characterized by low temperature, of course, and this has several important effects on quality. For example, the ability to dissolve oxygen to saturation dramatically changes with temperature (Figure 3) (Alter 1973a). Viscosity also varies with temperature —it is about 80% greater at 0°C than at 20°C. This makes quite a difference in the design of treatment facilities, and so does the effect on time of chemical reaction. It is not fully quantified, just yet, but it is becoming clear that chlorine or other disinfectant is slower to react, and some organisms harder to destroy, in cold water.

Oddly enough, the opposite problem is also common in the Arctic. The vital need to keep water from freezing in the pipes requires that the water or the pipes be confined to heated "utilidor" tunnels. This is rarely done just right, for to be on the safe side, the heating is often overdone. In the case of utilidors containing heating pipes, there may even be cases of hot water only available at the users' end of the pipe. Hot water causes treatment problems of its own.

Collection, Storage, Distribution and Thawing

Gathering Water

Gathering liquid water is not greatly different in the Arctic than in temperature zones. The problems concern (1) avoiding freezing, and (2) protecting pipes and pumps against ice damage. Where there are thaw bulbs, simple well-points into the thawed aquifer suffice. But the common structure built vertically over such a well is vulnerable to the shifting of surface ice during breakup and may well be destroyed as ice is swept away in the spring freshet. At river locations the buildup of overflow icings may submerge the intake structure in solid ice, not only precluding operations but also making repairs extremely costly and difficult. Nor are lake installations invulnerable. Figures 4 and 5 show

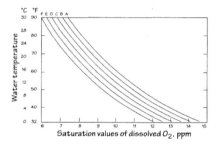

Figure 3. Oxygen solubility with temperature. Water exposed to air can dissolve oxygen in varying quantities governed by the degree of exposure, pressure, and temperature. Curves A through F are for altitudes in feet above sea level (A), 1000 (B), 2000 (C), 3000 (D), 4000 (E), and 5000 (F).

4 For example, the piles of ice stored by some dwellings in some treeless northern communities become marking-posts for the canine residents, and the oft-repeated warning, "Don't eat yellow snow," has much validity. The advent of snowmachines has reduced the severity of this problem.

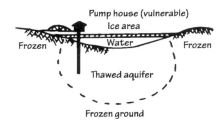

Figure 4. Wells may sometimes be established in the thawed ground beneath rivers or lakes. Through-the-ice completions may be vulnerable to ice forces.

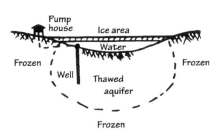

Figure 5. Under-ice completions may be required to avoid damage from ice (see also Figure 10).

the situation and one type of solution to the difficulty. Even where there is fine material of low permeability, unsuited for well-points or other underground completions, intakes have to be placed below the lowest depth to which ice may be expected to form, and the water line should remain below the level of danger from ice. Pipes may be buried in the beach, perhaps, continuing underground to the pump house, well away from danger of destruction or inundation by ice. Alternatively, a gravel island may serve as both an artificial aquifer and a mechanical barrier against ice. In that case, the gravel should be thick enough to raise vulnerable structures above the danger level, and to remain unfrozen in the collection zone. Insulation buried in the gravel may be required to retard frost penetration into the aquifer.

Temporary or semipermanent intakes are sometimes contrived by simply cutting a hole in the surface ice and dipping or pumping water from the hole. Such intakes require continuous attention but are suitable for small installations where water is hauled in batches to the point of use. During the construction and operation of the Distant Early Warning Line (DEW Line) of radar stations, it was common to use tracked vehicles to pull sleds fitted with pumps and water tanks from the sources to the sinks. The driver would reopen the hole through the ice and pump his tanks full. When the lake water was reduced to smelly dregs, he would seek another lake, and so on. Such water supply techniques, while perfectly possible, resulted in a situation where "the cat skinners haul water for the cooks and the cooks cook for the cat skinners"—and a camp of many existed to support a few operators and technicians.[5]

Storage

Even tropical developments require water storage, usually in elevated reservoirs, to provide for a fluctuating use rate and ready availability in case of fire. In the Arctic, the storage volume needed is likely to be much greater: first, because the stored water may be the only liquid water available for much of the year for fighting fires, and second, because the source may exist in convenient form only during part of the year.

In the first permanent camp for the British Petroleum Company near Prudhoe Bay, for instance, it was decided to provide for a volume of storage sufficient to supply water throughout all the freezing months. Pumping from the nearby lake is therefore confined to summertime, "when the livin' is easy." Not only that, but part of the storage is in the form of a swimming pool, where it serves a useful purpose yet remains instantly available in case of fires to fight—a brilliant way to rationalize luxury in the North. Storage in "natural" reservoirs is not so common, yet both in America and in the Russia there are places where man has made reservoirs by damming streams. In Alaska, there are dams on permafrost sites at Barrow, at Kotzebue, and formerly at Livengood. There will be many more as time passes, as there are already in the Siberian North, and construction of these reservoirs involves the solution of many special problems. But the problems are solvable with a little exercise of knowledge and ingenuity.

5 A chief engineer of construction on the DEW Line once told me that chartering aircraft to haul water from Fairbanks to the Arctic would have been cheaper than exploiting the local water this way.

Even storage in insulated tanks has some problems. In a legendary case, a shell of ice grew at the top and sides of a large cylindrical tank on Alaska's west coast. The unfrozen "core" of water was withdrawn through piping and valves near the center of the tank. As water was used, the support of the vast ice sheet that had formed at the water surface was lost. The horizontal sheet was then supported only by the cylinder of ice around the periphery. The heavy ice plate eventually broke and plunged down, destroying the control pipes as well as the access ladders needed for the tank's repair. It takes a lot of experience (hindsight) and foresight to anticipate and forestall such an event as this.

. . . And Conveying It . . .

If there is a continuous flow, and if the source can be maintained a degree or two warmer than the freezing point, it may be possible to convey water a considerable distance in an insulated pipe without the addition of auxiliary heat. It is simple to calculate fairly accurately the heat loss from an insulated pipe—or, conversely, to determine how much insulation is required to reduce heat loss to a tolerable level—and this loss can be supplied by flow of water warmer than 0°C, or by friction, or by some sort of "heat tracer." Heat tracers can be electric resistance wires or tubes filled with heatable fluid, placed adjacent to the pipe within the protective jacket of insulation.

It is wise in most cases to provide for heat tracers whether or not the flow is supposed to be continuous, for there can be accidental blockages (power outages, frozen pumps, whatever) that stop the flow and render the motionless water in the pipes vulnerable to frost. And a pipe once solidly frozen may have to stay that way until summer comes, unless the designer has planned ahead.[6]

Heat loss through cylindrical shells is governed by the equation:

$$Q = \frac{2\pi(T_2 - T_1)}{\ln\left(\frac{r_2}{r_1}\right)}$$

where Q is the rate of heat flow per unit length of pipe, T is temperature at the inner or outer surface of the shell, and r is the radius to the shell surface, as shown in the drawing below.

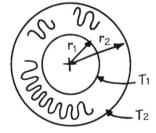

It is often desirable to put a thermostat control somewhere on the tracer that will turn off the heat when the pipe is in no danger of freezing, and to turn it on in times of need. If the time of need occurs due to a power outage, the tracer will likely be inoperative until the power comes on again. But no matter; when the power is restored the pipe will thaw satisfactorily—if it hasn't broken in the meantime.[7]

6 Some villages use a temporary, above-ground system of uninsulated pipes during the summer. The system is dismantled during winter.

7 Some tell a true story about one work-man who carefully placed the thermostat controlling heat pipes in a major housing development outside the pipe insulation. The insulation was combustible.

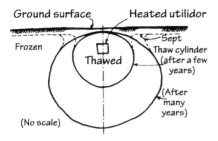

Figure 6. Any heated conduit, however well-insulated, will create a thaw cylinder in frozen ground. Insulation will slow the rate of its formation and reduce the extent somewhat. (Adapted from Lachenbruch, 1970.)

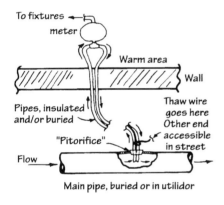

Figure 7. House services my be kept thawed by lavish use of insulation plus continuous circulation induced by the velocity of water in their main pipe. If the house service were to freeze despite these precautions, it may be thawed by passing high currents of electricity through the service pipes to an electrical connection located at the street end,

Distribution: Utilidors

A common way to handle pipes for distribution of water is to place them in utility tunnels, or utilidors. These passages are kept above freezing and they may also carry heat distribution lines, power lines, sewers, and people. There are some distinct advantages to utilidors. For one thing, heat that must be provided to protect water lines also protects the sewers. For another, heat inevitably lost from steam or hot-water pipes keeps the utilidor warm. Moreover, if the utilidor is large, there can be access for repair and alteration to pipes inside, away from the frigid weather. Further, astute designers may sometimes allow corridors and utilities to coexist in the selfsame structure, to the mutual advantage of all and with a net saving of energy as a bonus.

In areas where the ground is rich in ice, utilidors are commonly placed above ground; otherwise the problem of burying a warm conduit in the ground is difficult and costly to solve (Figure 6) (Lachenbruch, 1970). Where utilidors must serve individual dwellings or other small buildings, diminutive tributary utilidors ("utiliducts") may be used.

Nearly all modern communities in the North have some form of utilidors, and this will be the trend for the foreseeable future. Existing towns, however, are not laid out in such a way as to make utilidors practical. It will take a great deal of intelligent effort to adapt such towns to running water and sewer systems.

Some towns (Fairbanks, for example, and Unalakleet and Yellowknife) are not located over ice-rich ground. It is therefore possible to prevent distribution pipes from freezing merely by circulating water. The water system is formed by intersecting closed loops, with water pumped through every section of pipe for all normal conditions of use. Provision is made for heating the water at some point in the system, if required. In Fairbanks, the water is warmed (by "waste heat") to a temperature suitable for chemical treatment, then it is circulated through the lines. Figure 7 shows how house services are handled. A circulating system is much cheaper than a heated utilidor and allows somewhat more freedom in civic design.

Thaw the Pipes Harmlessly

Pipes will freeze from time to time as one thing or another goes wrong. But they have to be thawed to be returned to service. Many dwellings are destroyed each year by fires started because of overenthusiastic thawing. Usually this is because the first thought is of blowtorches, weed burners, and other flaming heaters. It's better, and almost as fast, to blow warm air against the pipe where ice is suspected. Some vacuum cleaners can be arranged to do this. (Blow warm interior air only. Don't expect the machine, which is merely an air pump, to thaw pipes with air at forty below). Hair dryers are ideal.

For thawing underground pipes, it's a good idea to arrange dual pipes or a pipe with a "thaw wire" so that high amperage from an electric welding machine can be sent through to thaw the inaccessible pipe quickly (Figure 8, page 72). As an alternative, a heat tape (electric resistance-heating wire) can be buried with the pipe. It may be either

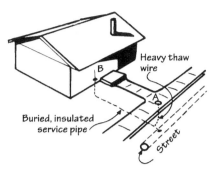

Figure 8. A generator may be placed between A, in the street, and B, in the house. (A welding machine is ideal because it can produce high amperage.) The service pipe will be heated enough to thaw. It is considered poor practice to continue this process so long that the plastic insulation or the solder joints melt away.

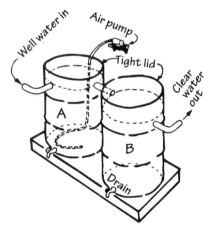

Figure 9. In this de-ionizer, a pair of drums are burned clean and painted with waterproofing paint. Tank A is aerated (perhaps with a dime-store aquarium aerator) and Tank B serves as a stilling basin, where iron oxides may precipitate and settle out. From time to time, the ketchup-like sludge must be drained away. Clear water is drawn from the tank just below the surface. Tight lids help control vapor and prevent contamination. Such a system is suitable for low flow rates.

permanently controlled by a thermostat or else arranged to be easily energized if the pipe becomes frozen. In the latter case, some sort of signal or circuit breaker might be included to protect against the possibility that the heat tape might remain hazardously hot too long.

Plastic pipe is hard to thaw. A nonconductor, it cannot be resistance-thawed with electricity. If it is thermoplastic, it should not be allowed to come into contact with hot heat tapes. And of course, open flame is out for thawing plastic pipes.

All these difficulties dictate that much careful thought is required to avoid freezing problems with conduits of plastic. However, tests by Shannon and Wilson, Inc., in cooperation with the Fairbanks Municipal Utility System, have shown that high-density polyethylene pipe (HDPE) can withstand steam thawing and even prolonged steam heating at normal steam thawing temperatures and pressures without damage.

Treatment

Provided that water is warmed to conventional, temperate-zone levels, its treatment is no different than that used elsewhere. It can be clarified, softened, de-ionized, aerated, chlorinated, or otherwise handled just as it is in other, more clement latitudes. And that is the way it is usually handled: rather than invent a new arctic technique, it is easier (and much more sure) to ensure the water a temperate environment and to treat it normally. Therefore most arctic water supplies, if treated at all, are treated by conventional techniques within heated buildings. It isn't cheap to do that, but neither is any likely alternative. So, sadly, we will expect few breakthroughs in the near future.

For small systems—say, of individual dwellings—there is need for cheap adequate iron-removal systems, because the easily obtained ion-exchange-type softeners are rarely adequate to handle the high concentrations of mineral and organic iron that sometimes occur in arctic water.

With ingenuity and a little plumbing, however, the harried householder may arrange a pair of open tanks, one equipped for aeration and the other for settling. With brown sludge of iron hydroxides, even a fairly simple system can go a long way toward reducing the iron problems (Figure 9).

How about Recycling?

Reuse of waste water for drinking, bathing, dishwashing and cooking is possible, but as yet not practical. ("Thank God," say some.) But it is entirely practical to save some water, the "gray water" from laundry and showers, for instance, to use for toilet flushing, fire fighting, vehicle washing, and such.

Where "new" water is in short supply, or where growth has outpaced the existing water system, water can be conserved to an extent by such recycling. The hitch in this practice is that two parallel piping systems have to be used, with attendant possibilities for a "cross connection"—the source of many a violent epidemic in the past and a source of recurring terror to engineers and public health officials. Nevertheless, recycling is a distinct possibility for the future. It can and must be safely done.[8]

Denouement

People generally have evaded rather than solved arctic water problems. In some cases we have simply enclosed our treatment systems and kept the pipes warm so that we face only the familiar problems of warmer climates. This is not necessarily bad: we've done much the same with people, with such success that the indigenous inhabitants of the North prefer the temperate zone living to the harsher way of life they formerly endured. But it goes a little easier on the user and on the environment if the designing engineer is familiar with what characteristics the Arctic climate will bring to bear, for good or ill, to affect his projects.

References

Alter, A. J. *Cold Water Quality*. Report IWR-39, University of Alaska Fairbanks: Institute of Water Resources, 1973a.

Alter, A. J. *Water Supply and Waste Disposal Concepts Applicable in Permafrost Regions.Permafrost: the North American Contribution to the 2nd International Conference*, Washington, D.C.: National Academy of Sciences, 1973b.

Alter, A. J., and J. B. Cohen. *Cold Regions Water Storage Practice*. Public Works 100, October 1969.

Alter, A. J. *Water Supply in Cold Regions*. Cold Regions Science and Engineering Monograph 111-C5a, Hanover, N.H.: U.S. Army Cold Regions Research and Engineering Lab., 1969.

Alter, A. J. A Guide to Arctic Water Works Thermology. *Consulting Engineer*, April 1957.

Anon. What Happens to a Well Casing as Permafrost Melts. *Alaska Construction and Oil* 11(8), 1970.

Clark, Lloyd K., and A. J. Alter. *Water Supply in Arctic Areas: Design Features*. Proceedings of the American Society of Civil Engineers, *Sanitary Engineering Journal*, Paper 931, 1956.

Cohen, J. B. *Water Supply in Alaska*. Proceedings of the 2nd International Symposium on Circumpolar Health, Oslo, Norway, 1971.

Hoffman, R. C. *Above-ground Utilidor Piping Systems for Cold-weather Regions*. Technical Report R-734, Port Hueneme, California: Naval Civil Engineering Lab., 1971.

Lachenbruch, A. H. *Some Estimates of the Thermal Effects of a Heated Pipeline in Permafrost*. Geological Survey Circular No. 632, Washington, 1970.

Peyton, H. R., P. R. Johnson, and C. E. Behlke. *Saline Conversion and Ice Structures from Artificially Grown Sea-Ice*. Report IWR-1; University of Alaska Fairbanks, Institute of Water Resources, September 1967.

Tilsworth, T., and F. J. Damron. *Wastewater Treatment at a North Slope Industrial Camp*. Fourth Joint Chemical Engineering Conference, Vancouver, B.C., Canada, 9–12 September 1973.

Williams, J. R. *Ground Water in the Permafrost Regions of Alaska*. U.S. Geological Survey Professional Paper 696, U.S. Government Printing Office, 1970.

8 More than 60 years ago, Professor Imhoff observed that water was taken out, treated, used, and returned to Germany's Ruhr River in amounts more than seven times the flow of the river at any point. This sort of thing is now common throughout the developed countries. That's real recycling—but the casual citizen is not alarmed by it.

Waste Management in the North

CHAPTER 8

Mary, Mary, quite contrary,
Why won't your sewer flow?
With pipes all stuck
You need Bud's truck
To thaw them down below
    —*Book of Nursery Rhymes* by Bud
    Hilton, a.k.a. the Underground Poet
    of Fairbanks

If You Think You've Got Problems

Introduction

What makes the Arctic so special regarding waste disposal, among other things, is its remote, frigid character: distances and costs are often so great as to discourage recycling most wastes, yet the weather is too cold for too long for nature unassisted to reduce wastes to safe and inoffensive residues within a reasonable time. Solutions that are complex and expensive enough in lower latitudes become extremely difficult in the Arctic. Most vexing and harmful to humans is the sanitary waste disposal and treatment dilemma.

Until comparatively recently, waste disposal was not considered a particular problem in the Arctic, except of course by those who live there. This was particularly true in the rural portions of Alaska, which are off the road system, remote from conveniences, and sometimes insultingly out of mind. Now, however, as the populations of remote communities continue to increase, there is an awakening world-wide consciousness of the need to both preserve the environment and to provide for more reasonable levels of public health. The clumping of population, in addition to the inflow of goods from developed areas, creates aggregations of trash and human wastes at the very moment when local and even national concerns for ecological balance and environmental health are flourishing. Jet-age aspirations for an easier, more abundant life are emerging in a frontier polar setting. This results in a confrontation exemplified by many inadequate attempts to reduce arctic waste disposal problems. As we shall see, contrasts and contradictions inevitably arise.

The law, for example, is for the first time concerning itself with the problem by attempting to enforce nationwide standards (the same standards that challenge engineers in more populous lower latitudes), causing considerable consternation when what works everywhere else fails miserably in the Arctic. For not everything is known about rendering wastes acceptably invisible and harmless on the permafrost soils, under the snows, and in the cold waters of the Arctic. We know that pathogens tend to persist, and organic wastes decompose slowly in chilly environments. Also, trash accumulated in dumps or "sanitary landfills" becomes permanently frozen and is likely to be preserved intact for the amazement and amusement of future archaeologists.[1] To compound the problem, incineration of trash, landfilling of garbage, and treatment and disposal of human wastes may create significant secondary pollution.

We must be ultracareful lest unintended results of our efforts are interpreted by future generations as environmental and societal

---

1 ...who may well come to the conclusion that the people of the "Snow-go Culture" worshiped body wastes, since they were preserved so carefully and so tidily in the permafrost.

carelessness. Surely, no other pioneering industry has ever before been scrutinized by so fervent a grove of environmental and amateur watchdogs as has the oil industry on Alaska's North Slope. As a result, a high quality of living has been created in spite of obvious hardship and extreme conditions.

Today, we must scrutinize ourselves and our responses to the rural sanitation dilemma, solid waste disposal, and supply of safe water and be equally proactive in solving these problems. Happily, most of the problems we face have technical solutions that are thoroughly understood. We know what we can do, now we must learn *how to do it* in sustainable and affordable ways that do not sacrifice quality and do accommodate local lifestyles.

In Alaska, there are endless examples of extremes of seemingly all descriptions. This includes a wide range of sanitary waste disposal methods, ranging from "honey buckets" (a common Arctic euphemism for a container of raw sewage) to sophisticated wastewater collection and treatment systems. Any one who has had the honor of dealing with any aspect of a honey bucket, particularly in an Arctic household without running water, will readily understand why this is considered to be the most rudimentary of methods.

Even outside privies and outhouses may be preferred over honey buckets, which must be hauled and emptied year round under intolerable conditions. At least the trip to the outhouse is usually empty handed, however cold and unpleasant. No wonder we yearn for sophisticated systems that seem invisible and trouble free to users elsewhere.

The Basics

The Privy

The pit privy is about as primitive a method of collection and disposal as most people wish to tolerate. Although users sometimes turn an outbuilding into an architectural wonder in the wilderness, it is usually an unheated little house placed over a pit big enough to hold the expected accumulations of a year or more. The fact that a pit privy is at once the beginning and end of the waste disposal process is one of the things it has going for it: when the pit is full, the structure, throne and all, is moved over a fresh pit and the old one is covered over with dirt and abandoned. (Alternatively, if not frozen, the pit can be pumped out and reused). Human nature often rebels at the use of an icy outhouse, so the chamber pot ("thunder mug") has its place beneath the bed. It requires frequent emptying into a privy pit.[2]

In most parts of the Lower 48, a privy might remain over the same pit for years, its accumulations moderated by aerobic decomposition in the earth and by drainage. In the Arctic, where the ground is frozen most of the year, if not permanently, very little decomposition takes place and the pit fills steadily. Indeed, north of the Brooks Range, the labor involved in digging any appropriate pit is prohibitive. Many opt, instead, for honey buckets.

2 The artful use of a thunder mug can extend the winter life of a pit privy, which otherwise would develop a spectacularly sharp stalagmite growing exactly and inexorably upward toward the center of the hole.

Figure 1. The outdoor arctic privy. A privy "with a view" can be devised to allow easy collection. Unheated, this privy discourages loitering in the winter.

Figure 2. The inside outhouse. An external door permits unobtrusive scavenging and replacement of honey buckets.

The Honey Bucket

In some northern communities, privies are built high enough to accommodate a container (often an oil drum) that can be traded periodically for an empty one. A modification of this system, used with a deodorizing chemical, may be enclosed within a dwelling, perhaps with an outside door to facilitate the exchange (see Figures 1 and 2). Nevertheless, there remains the problem of more or less frequent periodic disposal. However sophisticated the throne and its receiver may be, sooner or later the waste must be collected and carted away.

Honey buckets, under the best of circumstances, must be hauled and dumped frequently. This requires carrying the typical five-gallon pail with four to five gallons of sometimes sloshing raw human waste contained in a plastic bag, to the lagoon through arctic cold and wind, clouds of mosquitoes, and/or ankle deep mud, depending on the season. It's never a particularly pleasing task, and human nature being what it is, sometimes results in an abbreviated trip and somewhat indiscriminate placement of the waste somewhere short of the intended goal—and then there's the problem of finding a place to wash one's hands.

Getting It Together: The Collection System

From Isolated Homes

Householders living in arctic isolation find it possible to dispose of human wastes by dumping the containers far enough away so as to be inconspicuous. As long as the population is sparse, this method will work. A true believer in keeping the environment livable could well make some rules about disposing of honey bucket wastes:

1. Make a box or trench above the water table (or permafrost table) so that the liquid parts can evaporate or drain away.
2. Allow feces, food scraps, and other organic detritus to decompose together in the box over time (rather than dumping each load in a new place).
3. Arrange for a tight lid to discourage flies, dogs and wildlife.

After the waste is "seasoned"—when it has decomposed and dried enough to be inoffensive—it may be used as fertilizer for flower gardens, lawns, or tundra. Or the material, if adequately dehydrated, can be burned. There is evidence that pathogens exposed to a long winter "in the dry" diminish in viability. Amoebas are definitely killed by freezing and perhaps some bacteria as well. However, little is known about the fate of viruses, however, so human wastes should not be used on vegetable or other crops meant for human consumption.

Community Haul Systems

A somewhat higher level of sanitary waste disposal is possible using honey buckets if someone is hired to do the dirty work of hauling them for an entire community. The euphemism used to describe this minor step forward is "community haul."

Community haul has long been an important factor in cities where sewers are absent. Los Angeles had honey wagons as late as 1910, and some famous cities of the world still do. To be sanitary, waste containers must be picked up, thawed, dumped, washed, and returned to service.

This theoretically reduces the level of an individual's contact with wastes and may reduce the total amount of spillage on the way to the lagoon. Notice the term "theoretically." Communities have tried variations on this idea, including centralized honey bucket collection points or bins (big honey buckets on wheels), which can be hauled by snowmobile, all-terrain vehicle, or some other vehicle driven by a person hired for the purpose. This method incurs costs for operation and maintenance, not to mention injury to the dignity of the hired hand. Would you do it? And be willing to fix the equipment when it broke down?

These are real costs for a slight improvement, if any, in service. Raw wastes come in intimate contact with the community, and we haven't really broken the link to endemic waste-borne disease. It doesn't really improve household aromas very much either!

Sewers in the Arctic

In temperate urban areas, underground pipes carry sewage away to some mass treatment system, and the majority of its users never have to think about their workings or maintenance.

In areas of continuous ice-rich permafrost, burying sewer lines is usually impractical. Instead, sewers are normally carried above ground in insulated conduits, or utilidors—see Figure 3.[3] Otherwise, part of the heat required to keep sewers from freezing may ultimately melt a thaw cylinder around the pipe, with messy and expensive consequences.

When designing a facility from scratch, as Inuvik was designed, one can accommodate the above-ground utilidor by building dwellings extra high, so the sewers can flow by gravity, and by arranging the streets to

3 Utilidors almost always include both water and sewer pipes, and some of them carry heating pipes and electric and telephone wires as well. Some are built as pedestrian ways, too. For more discussion, see Chapter 1, "Permafrost: Its Care and Feeding."

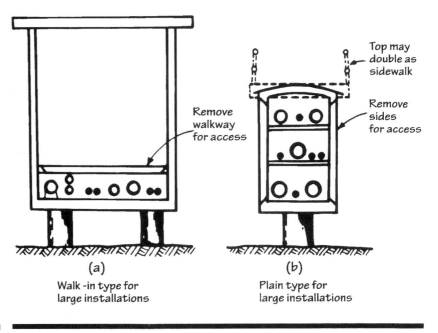

(a) Walk-in type for large installations

(b) Plain type for large installations

Remove walkway for access

Top may double as sidewalk

Remove sides for access

Figure 3. Above ground utilidors are needed where the ground is so rich in ice that melting would cause dangerous settling. This type can be difficult to accommodate to vehicular and pedestrian traffic, and must be placed carefully. They may carry electricity as well as steam, water or sewage. Conduits should be accessible at every point, and heat must be continuous in time and space.

circumvent the utilidors. However, existing camps and towns can rarely be refitted with above-ground utilidors, because these must cross roads or else require numerous lift stations. So attempts to provide sewers for towns not designed for them often require some pipe burial. If the ground is not ice-rich (that is, if it won't deform crazily when thawed), the installation can be designed in about the same way as in lower latitudes except that a combination of heat and insulation may be required to protect the pipes from freezing. And, as with all arctic water conduits, provision must be made for fast thawing and repair in the event of unexpected frost blockage.[4]

Generally, then, piped sewage systems include gravity, vacuum, and pressure sewage systems. Some thirty Alaska villages currently use piped sewage systems. For these villages, the predominant system in use is gravity flow, and some twenty villages are so equipped with accompanying lift stations and force mains. The villages of Emmonak and Noorvik operate vacuum sewer systems, but the system at Noorvik was significantly damaged due to freezing in 1990. Extremely cold weather, high ground/surface water, and marginal treatment facilities appear to be common problems associated with piped sewer systems in remote Alaska.

Roads are also necessary to support collection and delivery trucks, and winter snow removal is another factor that contributes to cost of operation. While capital costs are generally lower for truck-haul systems than for piped sewerage systems, operation and maintenance costs are often higher due to the cost of vehicle and road maintenance.

This complex system is all necessary to give arctic dwellers that emblem of civilization in the western world, the flush toilet.

The Standard Flush

To our shame, the standard flush-type toilet is used in the Arctic, where there is plenty of water. But even in areas where water is plentiful, the available water may be of unacceptable quality, present in unreliable quantity, intermittently thawed, and seasonally unavailable. Standard toilets use lots of water (ranging from less than one to more than four gallons per flush, depending on the ingenuity of their designer). They are, nevertheless, commonly used in arctic industrial camps, even though uneconomical, because, (1) money is not always the controlling element, and (2) people are, or are thought to be, most content with familiar fixtures. Select a water-flush, nonrecycling toilet requires a dependable supply of water and operative sewers, including some type of treatment process. Virtually all of the water used will, after all, become wastewater requiring disposal.

People who are unaware of the full cost of providing sanitary facilities in some arctic locations may still specify the familiar wasteful, standard, flush toilets, straining the pocketbook, the water supply system, and wastewater handling systems. The necessary complexity of such systems increases the probability of a malfunction serious enough to discommode the resident.[5] What's needed are simpler, less costly systems with fewer inherent opportunities for them to go wrong.

4 Electric or steam "heat tracers" should be adjacent to or in each pipe. This will prove to be cheaper in the long run than the traditional thawing technique: thrusting a steam hose into the icy sewer from an appropriate man-hole.

Use of the old four (gallon) flusher is pretty easy and "invisible" to the user, except perhaps when the time comes to paying for it. This of course involves installation of pipes and other fixtures, and usually requires a year-round source of carriage water available in each residence served. Initial capital costs are high, as are annual operation and maintenance costs. Who's to pay the bill and how?

Septic Tanks

Now we come to the standard septic tank, the traditional place that flush toilets empty into. With this system we can have flush toilets, and their use is limited only by the amount of water we are willing to haul and swoosh with.

But this familiar process also requires thawed and permeable soil for absorption of the septic tank effluent. Needless to say, this is sometimes impossible in permafrost areas. Septic tanks work fine, or at least acceptably, if some sort of leach field can be arranged and soil percolation can be achieved. The problem is that these tanks cannot be used everywhere but require careful evaluation of the conditions of use. This sort of evaluation, by the way, is required for all the various alternatives that one might consider for disposal of wastewater.

Septic tanks work by anaerobic decomposition, and low temperatures retard anaerobic processes, which are inherently slower to begin with. This creates an unfortunate problem because sewage production doesn't slow accordingly.

Septic tanks, commonly used for sewage disposal in temperate zones, work poorly in the Arctic: for one thing, their biochemical oxygen demand removal is low—around 35% at best—and the effluent is difficult to treat. Also, disposal of septic tank effluent in an absorption field requires permeable, thawed soil—a rare commodity in the land of permafrost (see Figure 4).

Septic tanks are seldom chosen for communities in the Arctic because our arctic soils are usually frozen and impermeable for a large part of the year, if not permanently. Even so, in warmer parts of Alaska, isolated rural residents find that septic tanks are adequate and that the

5 Government officials have closed down at least one facility because of an overloaded (and therefore ineffective) waste disposal system.

Figure 4. The household septic tank disposal method. This system depends on the use of enough warm water to keep the drain field from freezing and is not recommended for use over continuous permafrost. The septic tank is sealed against intrusion of air. The methane resulting from septic (anaerobic) decomposition must be vented away. The drain field should be buried uniformly, about 3 feet deep. This system will work satisfactorily in areas with climates as severe as that of Fairbanks, Alaska, if the ground is aerated and remains permeable. In areas farther north, where the climate is a little colder a little longer, this septic system will not function.

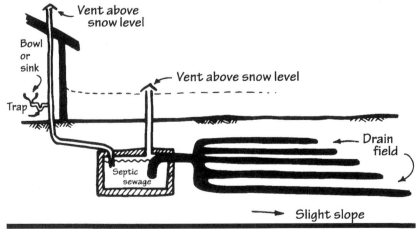

effluent from them is oxidized, during the short summers, in the upper, unsaturated portions of the active soil layer.

The problem occurs in winter when the customary drain fields freeze and effluent is backed up into mounds of ice and liquid near or on the soil surface or beneath the snow. This often results in an unpleasant environment, both atmospheric and underfoot. These conditions are short-lived and tend to become serious only during breakup, although a sensitive nose may be offended at any season. (Some people permit their septic tanks to drain into a private "facultative stabilization pond." This strategy, marginally successful in the Alaska interior, is not officially sanctioned.)

In any case, septic tanks must be emptied and cleaned frequently (at least every five or six years, and preferably every other year) even under the best conditions.[6]

Common problems with septic tanks include those associated with drain fields (frozen, too small, plugged, and otherwise failing); high groundwater; no provision for sludge disposal; poor soil percolation and bedrock problems; becoming plugged after a few years or are damaged by freezing conditions, overflow, and other failures; sewage appearing on the surface; raw sewage discharging into receiving waters; and poorly constructed manholes, transit pipes, and sewer mains. As you can see, it's not necessarily a pretty picture or a quick fix.

Go Easy on the Water: Low-volume Flush Toilets

So far, we've discussed ways that sanitary wastes can be collected and disposed of without the familiar swoosh of water and gravity flow, a significant issue, because in many places there will not always be enough water to flush with during certain parts of the year. This is especially true if all the water you get is what you are willing to personally carry. How often have you lugged five gallons of water from somewhere down the street and poured it down the second-story toilet just for the fun of it? Try it in the winter and write down for posterity the thoughts that come to mind!

There is a compromise, however, between "swoosh" and "no swoosh." Just a little water can make the flushing sound that in turn magically makes the waste go away. Enter the micro-flush toilet, of which there are several varieties. These toilets can be arranged to discharge into holding tanks, either large insulated tanks external to the home or smaller internal tanks. These tanks can then be periodically emptied by a pump or vacuum collection vehicle (with a tank of its own) operated by a community employee.

A short reasoning process expands this concept to include delivery of drinking water to the home. But it doesn't take a rocket scientist to realize that duplicate and carefully segregated equipment is desirable here! Operation and maintenance costs for such a system are significant, including the need to build and maintain roads (for example, snowplowing, repairing frost heaves, filling quagmires, etc.) and/or boardwalks for the haul vehicle to travel on.

6 Fairbanks has several well-to-do people who "pump and thaw" for a living. One such person used to advertise by eulogizing his services in iambic pentameter over the local radio.

Truck-haul Systems

The small vehicle haul system, a variation of the truck-haul system, is currently under consideration as a possible alternative to improve sanitation conditions in some villages. While similar in concept to the truck-haul system, there are significant differences. The wastewater holding tank is usually much smaller (100 gallons or less), and all-terrain vehicles (ATV) and snowmobiles employing small vacuum/pressure collection systems are used instead of trucks for sewage pumping, reducing the high operation and maintenance costs associated with the truck-haul system. This system does not typically include a water delivery system.

Currently, a prototype small-vehicle haul system is undergoing field-testing at Mekoryuk, Alaska. The system uses an eighty-gallon bladder as a holding tank located directly beneath the toilet. The tank is emptied by applying air pressure to the bladder to force the wastewater into the collection tank, which sits on a trailer towed by either an ATV or a snowmobile. Graywater is collected from the wash basin and used to operate the low-volume flush toilet. An example of this system under consideration is the Public Health Service "Crum-Kruzel" tank, which uses a heavy duty, 100-gallon polyurethane collection tank, a standard marine toilet, and a grinder pump to evacuate the waste from the holding tank to the collection vehicle.

A small but growing group of communities in remote Alaska operate truck-haul sewage systems. The typical cost per sewage pump-out is around $7.00, and the cost of delivered water (which complements the truck-haul sewage system) is near $.07 per gallon, but actual costs may vary widely (Iwamoto 1992). There are also monthly electrical costs incurred for heating the holding tank to prevent freeze up. Holding tanks are typically 500 to 1000 gallons in capacity, insulated with foam, and equipped with an electrical heater.

So now we have some idea about ways to reduce indiscriminate contact with sanitary wastes in the community, how to increase the amount of water available in the home, and how to improve prospects for washing one's hands. It is not surprising that to improve health we must break the connection between people and their sanitary wastes and provide water for washing. But what costs are we willing to support to do so? Certainly the cost for the systems discussed here are significant. Solutions acceptable to all will require some combination of new ideas, hybrid systems, and a willingness to be flexible.

Sewage Treatment and Toilets of the Future

Environmental and Other Technical Criteria

There are several criteria important to sanitation system planning, design, and construction. These include the following:

1. Geotechnical: It is key that water be available year round if one selects flush toilet systems (i.e., piped sewer systems, septic tanks, and truck-haul or holding tank systems) to provide the necessary volume for waste transport through the system. Piped water is normally

incorporated into piped sewerage projects because the two systems are mutually dependent. Piped water systems lead to increased water consumption, and then to a need for additional disposal capacity.

2. Acceptable topography: Gravity sewerage systems are dependent upon elevated terrain to achieve enough slope for gravity flow. Additional excavation and numerous force main installations are often cost-prohibitive in designing gravity sewer systems for villages located in flat terrain.

3. Permafrost conditions: The location, depth, and extent of permafrost (intermittent, discontinuous, or continuous) will have a direct effect on the selection of sewerage systems. Permafrost will impact buried piped sewerage systems, individual and community septic tanks, leach fields, and holding tanks.

4. Groundwater table: The groundwater table will also affect systems external to the dwelling (piped sewer systems, septic tanks or leach fields, and holding tanks). Groundwater infiltration may over-saturate leach fields, infiltrate piped sewer systems, and eject unballasted holding and septic tanks through buoyant flotation.

5. Area susceptibility to flooding: Similar to high groundwater levels, seasonal flood waters will adversely affect piped systems, septic tanks/leach fields, and holding tanks.

6. Temperature ranges: Temperature extremes can cause piped sewerage systems to freeze. Leach fields also are subject to freezing, and composting toilets and septic system operation can also be adversely affected by very cold temperatures. The latter generally return to normal with suitable temperatures.

7. Annual precipitation, snow accumulation: If one chooses a truck-haul or small-vehicle haul system, drifting snow and muddy roads may restrict access to homes needing service for significant periods of the year.

8. Soil conditions and stability, depth to bedrock: The construction of piped sewer systems, septic tanks and drain fields, and the installation of holding tanks will vary significantly based upon the local soil conditions and stability and the location of bedrock.

9. Readiness and Potential: These two intangible factors, while difficult to objectively quantify, can be useful in determining whether a village has demonstrated the necessary responsibility, motivation, and maturity to manage, operate, and maintain appropriate waste-disposal systems.

Another important factor that should be assessed is spending priority within each village household. Essential services such as heating oil, electricity, and food are quite obviously high-level priorities because they ensure survival. Nonessential expenses, which may nonetheless be perceived as very important, may also consume a large percentage of monthly income. This income might otherwise be committed to pay user fees for improved wastewater disposal systems, which would in turn improve health and well being.

It may be argued that the single most important feature of a community in determining its potential for success in developing municipal infrastructure is its ability to pay in perpetuity for all sorts of basic services due to all human beings. On the other hand, it may not be reasonable to expect exactly the same services found in the urban metropolises of Anchorage, Fairbanks, and Juneau. This is not to say that the same effective service shouldn't be provided, just that identical services may not be reasonable. Most people who live off the road system like living there for the very reason that it isn't like the cities! Many forget that Anchorage was the bush within living memory. At any rate, we must seek solutions that accomplish basic needs. These solutions will inevitably include sound and probably broad-based financing, and there are few communities that won't bear their equitable share.

During the last decade or so, there has been increasingly insistent clamor for running water and "automatic" waste disposal in rural villages. After all, goes the reasoning, if rural homes in Kansas have flush toilets, surely we deserve no less here in Alaska. This feeling is augmented (despite the argument that the technical, financial, and societal difficulties are enormous) by the observation that military and oil-related camps have plumbing, and so do hospitals, even when located on permanently frozen ground. Still, "modern" conveniences haven't invaded all rural Alaska yet. Many small villages still rely on unmodern systems that appear to serve them fairly well.[7] Yet some problems persist, and those often seem more dependent upon our ability and willingness to support costs, rather than on suitable technology.

Biological Treatment

Biological treatment is by far the most common and has the widest range of processes, from stabilization lagoons to highly sophisticated, proprietary, extended-aeration systems.

Aerobic Systems: Lagoons

The lagoon system is physically the simplest: wastewater is ponded in a shallow reservoir of suitable detention time, perhaps equal to the duration of the frozen period of the year. Various organisms, primarily the bacteria, can almost completely degrade (up to 95+% biochemical oxygen demand removal) organic wastes, but at rates that slow as temperatures drop. This is very good indeed. In the Arctic, however, there are long periods of ice cover, and sewage lagoons do not function so readily. The use of lagoons for raw sewage north of the Brooks Range in Alaska is not encouraged, although some users have reported satisfactory results for the southern slopes of that range. For a lagoon to work well in the Arctic, it is wise to expect septic (anaerobic) conditions during the period of ice cover, and one may also expect an interlude or two of offensive odors during the fall and spring as ice forms and breaks up.

Lagoons are not in and of themselves dangerous to public health if they can be isolated from the community. Too often, however, travellers in the rural part of the state find children playing uncomfortably close to, or sometimes in, suspect tundra juice. Simply building a fence is

7 Some towns that have been successful in getting community water supply and waste-disposal systems find that they require intolerably costly maintenance and repair and need to be extensively modified to work at all, for in the Arctic there are all the problems of lower latitudes plus many unanticipated by the designers.

insufficient. Fencing rises and sinks with the moving tundra, and dead level top rails are rare in permafrost country. Any lagoon can become a hazard without periodic monitoring and maintenance.

In the future, waste treatment will inevitably become a more and more important consideration, although safe and efficient systems may also inevitably increase the complexity and cost of municipal infrastructure.

Activated Sludge and Extended Aeration

Activated sludge and extended aeration processes are more effective than lagoons, and more expensive as well; they may require a collection system and generally must be housed and/or heated in the Arctic.

Systems for aerobically stabilizing sewage in minimal space are highly desirable but also costly, for they work best at comparatively high indoor temperatures. Fortunately, many manufacturers sell off-the-shelf equipment, and these, if competently supervised and maintained, can do a credible treatment.

Sewage is typically passed through a grit chamber and a comminutor (grinder), then sent directly into a tank where it intimately mixes with air. Once biological action starts, the process is more or less self-sustaining, as long as nutrients (sewage) continue to be introduced.

After aeration, the sewage is detained in a clarification tank to promote settling of solids. Sludge may be drawn off and recycled within the treatment process, with some being "wasted," perhaps to an incinerator, or, after dewatering, it can be disposed of as a solid waste.

The end products of hydrocarbon oxidation are water, carbon dioxide and mineral ash, with occasional gaseous sulfides and soluble nitrogen oxides. The heat released by this "wet burning" is the same as the combustion heat produced by any other means, and in carefully contrived systems it can significantly assist in producing the warm temperatures necessary for aerobic stabilization.

Other Aerobic Possibilities

There are variants of aerobic processes, such as the "oxidation ditch," and the trickling filter. However, these are susceptible to freezing and are unfavorable for this region. In general, methods of aerobic decomposition are well advanced and will undoubtedly be the first choice for waste treatment for some time to come.

Composting Toilets

Composting toilets have been used in Alaska communities for over twenty years. In 1971, humus biological/composting toilets were purchased as part of a Housing and Urban Development (HUD) project in Selawik. Early designs could not effectively handle liquid shock loading and frequently failed (which meant, in practical terms, that everytime a family held a party or large family gathering, the toilet died).

Fort Yukon experienced similar problems with humus toilets, and an informal survey in 1977 indicated that of the twenty-five units installed as part of a HUD project, eleven families had the units removed from their homes, one unit was not in use, three units were still in use, and the status of the remaining ten units was not determined. Common complaints noted were odor problems, overflow, inability of the units to handle liquid overloads, and the high electrical costs necessary to heat and evaporate accumulated liquids. Other villages that may have firsthand experience with composting toilets include Kivalina, Barrow, and Deering.

Current state-of-the-art composting toilets operate on the same principle of biological decomposition. Design improvements include compost pile agitators, timed heaters, improved ventilation systems (to reduce odors), and improved liquid evaporation systems. No formal field testing in individual homes has taken place. No validated results are available to determine how applicable composting toilets might be to remote Alaska communities. Potential concerns include the energy costs necessary to maintain the heater and evaporator/blower, and the willingness of people to accept compost toilets.

Less apparent reasons why composting toilets generally failed in Alaska villages include a lack of initial acceptance by homeowners and an inadequate user training system. In the past, some engineers neglected to consult the villagers on whether or not they wanted an improved waste disposal system. Excluding the community from participation in decisions that intimately affected their lives contributed to homeowner indifference to and ultimate rejection of the system. This may have occurred in the case of the composting toilet, where the units were thrust upon homeowners without their concurrence and with little or no training.

Combustion Toilets

Sludge or screenings from primary treatment can be burned in incinerators, as can the contents of honey buckets. Feces alone can produce just a little more heat than necessary to sustain combustion in a well-designed burner, but any liquids must first be drained away. Much the same can be said for sludges resulting from various treatment processes.

Where trash and petroleum wastes can be processed simultaneously with human wastes, burning can be arranged to evaporate liquid wastes as well. This makes incineration, under controlled conditions and at high temperatures, very desirable (except perhaps for its cost), since the final ash is a completely stabilized mineral material that can be safely deposited on land, perhaps as landfill material.

Gaseous residue can be dispersed innocuously by the arctic wind, unless atmospheric temperature inversions—a common occurrence in many Alaskan valleys—prevent this. Designers and users of such a system must find ways to avoid smoke or ice fog pollution of inhabited places nearby.

Hybrid Composting-graywater Treatment

Hybrid composting-graywater treatment systems are perhaps the best example of applied ingenuity to come down the slough in at least a few years. This system combines composting with the more or less standard extended aeration process used in countless wastewater treatment plants throughout the world. Capable of providing a reasonably well treated liquid waste stream and of accommodating excess liquids, this system is a great improvement on composting alone.

No method is problem-free, however, including hybrid systems. Although they cost less than some systems, installation is not cheap. They tend to be more mechanically complex than other less active systems, and with complexity comes an inevitable need for maintenance and operating attention on some periodic basis. Operating costs aren't insignificant, although they can be minimized with careful installation. A good design can decrease the money spent for heat energy through insulating properly and limiting heating to the parts of the process that need it most. Heating is an unavoidable concern in the Arctic.

In addition, all systems are to some degree regulated by various agencies, whether local, state, or federal. Regulatory agencies tend to be very sensitive about approving individual waste treatment systems, which is after all exactly what hybrid systems are. They fear an explosion of varied single-home systems, although they regularly approve single home septic systems that, in most places, seem to work just fine.

Individual hybrid-graywater systems are new and therefore suspect in the minds of many agencies. They would do well to remember that even the time-tested septic system often causes problems in densely populated areas and in areas where the soil drains poorly or where it drains too quickly near homes that depend on private wells for drinking water. Agencies face a real dilemma. These new systems can meet the needs of arctic communities, but regulators are reluctant to give them blanket approval. On one hand, solid and liquid effluents from hybrid systems may be just fine; on the other, they require close and consistent maintenance, and thrifty homeowners often succumb to the temptation to cut costs and figure out shortcuts that may lead to health problems for everyone around them. Given the right environment, septic systems require less watching and less annual maintenance.

In any event, the inherent value of combination systems and the great need for improvements over some of the rudimentary systems now used will force some solution or compromise on these issues. One of the most positive aspects of hybrid systems is that a good deal of thought has already gone into *both* wastewater and water issues. A reduction in waste-borne disease is likely, once clean water is easily accessed and wastewater effectively accommodated in a community.

Physical-chemical Treatment

When community wastes are fairly dilute, they may be separated from the water by the same basic processes with which drinking water is treated. Added chemicals produce a gelatinous floc that settles out easily, after which the water is filtered (perhaps through sand filters, and later, possibly through activated carbon to remove color or odor). Such plants can produce a fine, high-class effluent that may look "good enough to drink," but don't try it. Unfortunately, they are expensive to build and operate, and they require the attention of skilled workers. They are not for amateurs, nor for communities that cannot or will not pay for the associated costs.

A Word about Garbage, Trash, Metals, and Other Wastes

Unlovely metal wastes, characterized by the rusting skeletons of automobiles or industrial equipment and fuel drums, are not necessarily a health hazard, but neither are they permissible wilderness decorations. Hard enough to dispose of in industrialized areas, in the Arctic they pose a special problem: they rust away at an almost imperceptible rate, and their low value rarely encourages salvage for recycling.[8]

The solution to this problem will surely lie in the establishing of a few concentration camps for such junk, which will be accessible enough to justify the occasional barge or ship haul to a place where metal scrap can be reprocessed and marketed.

Burnables, of course, can be burned. This includes not only paper and wood products, but also food wastes and possibly sewage sludge. Yet the problem with burning is two-fold: first, it is expensive to provide an acceptable means of incineration (including the cost of collecting and handling the trash) and second, incineration can cause air pollution, particularly in regions subject to atmospheric inversions. Both problems can be lessened by incineration with provision for saving the heat generated, as is done in modern co-generation plants. Incineration, properly done, consumes organic materials with a minimum of side effects, including new carefully controlled processes that can cleanly burn wastewater sludges and honey bucket wastes with little in the way of obnoxious off-gasses. Byproducts of uncontrolled combustion are flue gases (mostly water and carbon dioxide), inorganic particulates from the stack (fly-ash) and possible unpleasant gasses. They represent the most noticeable (and therefore most likely to be criticized) elements. But in an area where fuel can be costly and scarce where heat is often desperately desired, and at a time when conservation of fuel is a recognized necessity, it must be the goal of the thoughtful engineer to use the heat available from trash. Burnable trash, for instance, has a heat content ranging from about 8,000 to over 13,000 BTU per pound of dry matter. That is not so different from coal, and it is better than some lignites (soft coals, like those found in Healy, Alaska) currently used for producing heat and electricity.

The problem in using trash for heat is an economic one, stemming from (1) the extremely variable rate at which wastes accumulate; (2) the inevitable fluctuations in the quality (i.e., heat content) of the wastes (which vary in water content, physical sizes, and shapes); and (3) the chemical variety, which ranges from easily consumed hydrocarbons like paper, wood, and oil residues to plastic products. Any of these may be contaminated by chemicals used in manufacturing and marketing the original product.

Faced with the high cost of using nonuniform materials destined to arrive at unpredictable times and rates, engineers are often forced to conclude that contriving and operating a plant that can minimize pollution and save heat is unreasonably difficult and expensive. They are often right. Even in large cities, where both the production rate and the quality of the wastes are fairly stable, the mere sorting out of the

8 Good-hearted but fuzzy-thinking antipollution officials have, in the past, been known to specify—and purchase—high-temperature incinerators capable of burning light steel scrap (tin cans, etc.). That was before they realized that the world's supply of fuel and iron was finite.

combustibles has proved intractable and costly. Nevertheless, ways are being developed to deal with these problems, and success is within the grasp of imaginative engineers. Meanwhile, householders continue to consume their dry garbage and combustible trash in stoves and fireplaces, thus accomplishing to a degree what the engineers still aspire to: disposal of trash together with a net saving of fuel.

Wet garbage includes the residue from food preparation and consumption (bones, parings, food scraps). It poses a special problem where there are no pigs or other livestock to feed and no handy sewers into which all will disappear (if it is finely ground and accompanied by enough water). In most areas of the Arctic, food scraps attract animals—gulls, foxes, wolves, ground squirrels, the occasional bear—and however romantic and fascinating this may be to the human inhabitants, the results can be tragic. Experts agree that feeding wild animals is dangerous both to the feeder and the fed—partly by bringing them too intimately together for human safety and partly by making the beasts less able or too numerous to survive in the wild. Adding a hungry bear, for example, to the excitement of honey bucket disposal isn't as romantic as it might seem.

Heritage and Swoosh: How to Pay for It All

And so, a pattern is forming: no carelessness will be tolerated in developing the Arctic, no streams will be degraded from their pristine purity, and no unsafe or unsightly disposal areas will be allowed and disease must be reduced. Alaska's Arctic (and Canada's too) will not be permitted to assume the unpretty and unhealthy look so long associated with "progress" of past decades. The image so far, with more to come later: the oil camps of the North Slope of Alaska are the cleanest, neatest construction camps ever seen—and the cost of keeping them so can be borne only by bonanza developments. The cash flow common to "progress" differs significantly from that of most arctic communities.

Subsistence, Local Economics, and Cash

One key issue in matching an appropriate alternative wastewater disposal system to a specific village is the ability of the user community to operate and maintain the system *in perpetuity* once it is built. This issue usually focuses attention on the long-term financial capability of the community, which in many cases may be very limited.

The importance of subsistence from an economic and sociocultural perspective is difficult to overemphasize. The sociocultural importance of subsistence is critical to the very existence of the rural arctic way of life, particularly for Native Alaskans. From early beginnings as hunters/gatherers, Alaska Natives have consistently relied upon the land to provide all their basic needs. In addition, religious and spiritual ties with nature have long been a part of the Native culture, and others should probably learn from this. While the introduction of Western culture has affected subsistence to some degree, the basic importance of this way of life is substantially the same as it was historically. Somehow, solutions must conform to cultural needs and should be developed with strong and effective input from the local people. It's becoming clear that the rest of us may have a lot to learn.

Subsistence is not a purely or even mostly economic concept, neither is it simply a minimal way of getting through the winter with body and soul intact. It is an important sociocultural pattern and is a dominant cultural symbol and practice throughout remote Alaska. People feel strongly about this and it provides significant bonds between family and community members. It defines domestic roles and personal identity, it represents great cultural achievement, it provides critical sustenance and commodities, and it enables great persistence through time and durability in the face of adverse conditions.

The subsistence lifestyle is an essential means of both physical and cultural survival. While fish wheels and modern technology have thrust Alaska Natives beyond historical subsistence and into a partial cash economy system, subsistence salmon fishing, for example, represents critical continuity with the cultural life of the past. In addition, the importance of subsistence frequently results in conflicts between periodic summer employment (cash earnings) and the need to maximize the salmon catch to survive the winter. Often, one finds oneself living in two very different and contradictory worlds.

As one might expect, career opportunities in technology fields are sparse in bush Alaska, where most communities must be reached by air or water. People who pursue such a career must leave their families and communities for a more urban area, in Fairbanks or Anchorage if one is lucky, or in the Lower 48 if one isn't. Thus supplying operation and maintenance for a waste-disposal system is virtually always dependent upon making funding from outside the community available to pay trained individuals to operate and maintain systems, to provide for repair and replacement costs, and to bear additional costs for fuel, heating, electricity.

Funding can be provided by several means. A monthly service fee per household is most common. Village Council funding is another method that has been used, but eventually their funds may be exhausted, which inevitably results in system degradation and/or failure due to diminishing resources. Federal, state, and regional corporation funding assistance may not be generally available to subsidize village operations and maintenance costs. Therefore, despite limited cash flow in most rural communities, the preferred option for maintaining sanitation systems is often through collection of user fees.

The success of any cost recovery system depends on collecting payments on a timely basis. In a village of fifth to one hundred people, it can be difficult to collect cash payment for services, sometimes because of poor management, accounting, and billing practices and sometimes due to a reluctance to enforce payment by denying services. Imagine living in an isolated community and having to collect money from your elderly parents or relatives or in effect forcing them to hand-carry their own wastes in midwinter because they couldn't pay the bill. In such a situation, a bill may go uncollected as a simple matter of human compassion.

While physical aspects of waste management are of considerable importance to sanitation engineers, an equal if not more important concern is the long-term economic health of particular villages. This is especially true considering the need for a community to sustain water and wastewater systems through effective financial management. The need for sanitary services in the village will exist (but may change in character) for as long as the village exists that is, in perpetuity.

Throughout Alaska, the level of sewer service can be linked directly to annual average per capita income. Of the 223 villages surveyed during the 1990 U.S. Census, about one hundred villages were using honey buckets. Among these, eighty-five villages (89%) fell below the average per-capita income, indicating that, aside from geotechnical constraints, economic conditions in remote Alaska villages may have a direct bearing on the ability of a village to support and maintain modern systems. It is not surprising that due in part to different economic limitations, a substantial amount of support must be found outside the local community rather than within it.

Several important factors affect village lifestyle, including village accessibility and the local importance placed on a subsistence lifestyle. An important consideration in planning for residential sanitation facilities is the number of persons residing in each household. Rural Alaska household size averages about four persons, not including extended families. Large effective household size poses additional problems in finding suitable sanitation system alternatives that have sufficient capacity and are still affordable.

Economic and sociocultural conditions in the rural Arctic often represent significant barriers to planning sanitation improvements for the various villages. From an economic perspective, the lack of a viable economy reflects the inability of a village to satisfactorily support an improved sanitation system. The lessons learned from previous system failures have clearly indicated the need for a strong financial management base within each village to ensure proper system operation and maintenance. Similarly, a firm understanding of sociocultural patterns is necessary to plan, install, and operate improved sanitation systems. Also, failure to involve the village populace from the outset may lead to system failure due to nonacceptance. It's clear that we must find ways to minimize waste in the Arctic, and that engineers and consultants don't have all of the answers. Local people must participate in arriving at creative, workable solutions.

Public Health in the Arctic

In the arctic past, defecation while on the trail took place in a tent or snow house for protection from cold winds and eager dogs. But in camps or villages, a chamber pot helped to preserve privacy and prevent frostbite.

So long as life was nomadic and people dispersed, such simple waste handling caused only minimal problems for public health.

However, once villages became stabilized by store, church, and school, this changed: human wastes became too prevalent and too noisome to ignore. Further, dark predictions of health problems became realities; there was soon evidence confirming that waste-related health problems really did occur.

Liquid wastes (graywater or nonsanitary wash water and the like) are especially difficult to handle when there are no sewers. They can be too voluminous for convenient disposal by honey buckets, and they often don't *seem* particularly nasty, which leads to improper disposal on occasion. This inevitably leads to environmental pollution and degradation of personal hygiene as well as of public health, since pathogenic organisms or other contaminants generally are present even in graywater. Unfortunately it often seems the householder's only option is to dispose of such water by flinging it out the back door in what seems to be the best direction (downwind!). People who design infrastructure for camps and villages, which are becoming more populous and increasingly scrutinized, must take care that such wastes pass through suitable collection and treatment facilities before their ultimate disposal. Sometimes wastes, particularly nonsanitary graywater, can first be recycled for use in flushing or for fire fighting.

As a response to these new threats of disease, in the early 1950s, Alaska (then a territory) took an official look at the prevalent casual methods of waste disposal and began to train village sanitation aides. These individuals, selected by their own neighbors, learned how to dispose of wastes without also contaminating water and land and how to develop safe drinking water sources with materials at hand. The value of the program ultimately depended on the personality, commitment, and competence of each sanitarian and the cooperation of village residents. Some villages reacted with pride and a general clean-up resulted. Others were unimpressed with the need (which, after all, was perceived mostly by outsiders). In these places changes were hardly noticeable. The program showed the greatest promise, however, and it was already showing signs of success when territorial funding was cut back. After several years, the village sanitation aide program was halted altogether, and the modest funds available were shortsightedly spent instead for facilities. These often proved more elaborate, more trouble-prone, and more costly than villagers could afford.

The level of sophistication of sanitary waste disposal systems in remote Alaska is highly variable among some 239 village communities. Systems range from modern pressurized potable water and gravity sewerage to the most rudimentary honey bucket systems.

Conditions in villages that operate honeybucket haul systems vary from efficiently operated, well-managed systems to those where honey bucket wastes are essentially dumped indiscriminately. Under these worst-case conditions, the potential for disease is unacceptably high. Williams (1993) reports 200,000 physician/patient encounters annually in the Yukon-Kuskokwim Delta alone. Also, village aesthetics and quality of life are degraded by the absence of adequate sanitation.

If honey bucket systems are to be eliminated, sanitary waste disposal alternatives that serve the public needs of the community must be identified. It only makes sense that, for long-term success, the selected alternatives must be:

1. acceptable to the users and chosen—at least in part—by them,
2. economically supportable on a perpetual basis, and
3. compatible with the lifestyle of the community.

Comments and Conclusions

Villages have traditionally been located adjacent to water. Rivers handily carry away unwanted wastes in summer, and in winter their ice offers convenient storage for waste containers (even on sea or lake ice). The community can expect that breakup will cause dispersal and sufficient dilution to prevent health hazards or aesthetic problems. This technique has worked, too, except for those years when the ice rotted in place instead of being swept away during breakup. There is now evidence that bacterial contamination is indeed carried long distances downstream, so much so that "out of sight, out of mind" disposal is no longer acceptable. There must be reasonable alternatives.

Collection of trash and garbage also requires system planning and scheduling. In the North, honey buckets must be picked up, thawed, washed, and returned to service. This is not a bad system, so long as the collections are prompt and sanitary and sufficient wash water is available and disposable. Barrow, Kotzebue, and Inuvik have all used scavenging systems in recent years.

Unfortunately, it is difficult to divert people from their desire for standard plumbing, however costly and wasteful it may be, and it is difficult to introduce new alternatives that achieve similar results. So the question arises: can one achieve success and happiness with conventional or alternative sewerage in the Arctic? The answer is, "of course." Engineers can do almost anything if people are willing to pay all the costs and accept all the consequences of using the installed systems.

Most waste treatment systems—particularly biological and physical-chemical—are neither simple or automatic. They require regular and continuous care and can be costly. Usually, people deal with the arctic waste treatment problem by avoiding it. They typically move a proven temperate-zone technique indoors. This usually turns out to be expensive, but given informed care and operation, it can almost always be made to work. With some attention to protecting conduits from freezing as they pass from building to building, and with provision for draining the systems completely in case of heating failure, there is nothing wrong with adapting temperate climate solutions to arctic problems. Most designers rely heavily on this technique.

There appears to be a direct correlation between the amount of clean, readily available water for a community's honey bucket system and disease incidence. In communities where people haul water by hand from a watering point or other source, honey buckets (either individually hauled or community serviced) tend to be the predominant waste-water disposal system. The lack of water is a key factor in choosing not to install either a holding tank/septic tank system or a piped sewer system. Correspondingly, if trucked or piped water is available, the level of sewer service will also rise and human contact with raw sewage decrease. Available clean water in sufficient quantities seems to be essential to maintenance of good community health, given prevailing methods of sanitary waste disposal.

Treatment and disposal are fundamentally different and distinct issues, although obviously directly related. Disposal should be dealt with first because waste removal is more critical to health and quality of life than treatment is. Treatment, however, needs to be dealt with too. Each community will have to identify its own specific needs and engineering liabilities and develop a system accordingly.

Alaska needs to develop a comprehensive statewide strategy for community-specific sanitation by finding a comprehensive strategy for each village based on local needs. All appropriate alternative and innovative systems should be evaluated to define their possible role as partial solutions to sanitary waste disposal in remote locations. Additional data are required on the usefulness of composting toilets in remote Alaska villages. For example, previous experiences with composting toilets may be based on early models, which may have been improved upon in recent years.

The disposal of sanitary wastes in remote Alaska communities can be improved with existing technology. Successfully applying available technology is a challenge that encompasses socioeconomics, public administration, and management issues. Acceptable solutions can be found if we make a comprehensive, interdisciplinary effort.

References

Iwamoto, L. A. *Planning Sanitation Improvements for the Calista Region: A Formidable Task,* Unpublished masters thesis, University of Alaska Anchorage, 1992.

Williams, Orie. Vice President of Yukon Kuskokwim Health Corporation (YKHC), Bethel Alaska. Personal communication, Alaska Science and Technology Foundation (ASTF) workshop, Anchorage, 1993.

Index

A
aerobic decomposition, 89
air ducts, 19, 20
air pollution, waste disposal and, 92
air starvation syndrome, 45
anaerobic decomposition, 84
Anchorage, 2, 3, 27, 88
anoxia, 45
aquifers, 70
Arctic architecture, 27
avalanche areas, 13

B
bacterial contamination, 97
Barrow, 3, 40, 73, 90, 97
basements, of igloos, 28
biological sewage treatment, 88–91
Boeing Jetliners, entryways, 28
British Petroleum Company, 70, 73
builders, speculative, 29–30
building materials, 28
buildings, xi, 14–18, 57
    see also dwellings
burnable trash, heat content of, 92

C
capital investment costs, 29
ceiling insulation, 31
central heating, 43
chamber pots, 80
circulating water system, 75
coal, 41, 47
coal-burning furnaces, 42
co-generation systems, 39
cold traps, 27
combustion toilets, 90
community haul systems, 81–82, 86
composting toilets, 89
conduit, heated, 75
convection heat transfer, 60
Corbusier, 16
cost recovery systems, 94
Crum-Kruzel tank, 86

D
dams on permafrost, 73
Deering, 90
de-ionizers, 76
demineralization of water, 71
desalination, 70, 71
Distant Early Warning (DEW) Line, 21, 73
doors, 32, 33, 35
drain fields, 84, 85
driveways, 14, 19

dwellings
elevated, xi, 15–18
orientation of, 13–14
snow houses, 27–28
temperate-zone *vs.* traditional, 11–12
water service to, 75
see also buildings

E
earthquakes, 18, 24
electric heating systems, 46–47
electricity, 39, 47
elevated buildings, xi, 15–18
elevation, temperature and, 13
emergency escape routes, x, 34
Emmonak, 83
energy conservation, 37–38
energy costs with various fuels, 39
entryways, 14, 28, 33
Eskimos, 37

F
Fairbanks, 3, 12, 13
fencing, 88–89
fiberglass insulation, 30, 32
fibrous insulations, 30, 31, 32
Fick's Law, 50–51
fire danger, ix–x
fire logs, 40
floor insulation, 31–32
floor temperatures, xi
flush toilets, 67, 83–84
foamed-in-place urethanes, 31
foamed plastic insulation, 31, 32
foam insulations, 30, 32
foil-backed insulation, 31, 57
forced-air heating systems, 42, 44, 46
Fort Yukon, 89
foundations, 15, 18–19, 20–25
freeze back, 9, 21
friction-fit batts of insulation, 31, 57
frost heave, 4
frost-jacking, 22, 23
furnaces, x, 42–47

G
global warming, 20
gravel pads, 24
gravity-air heating systems, 42, 46
gravity flow sewage systems, 83, 87
graywater, 76, 86, 90–91, 96
grease vapor, 50
groundwater, 68, 71, 87
gun-type oil burners, 41, 43–44

H
haul systems for waste disposal, 81–82, 86
heat distribution, 45–47
heat exchangers, 42, 45, 47
heat flow, 2, 30, 32, 60
heat gain, 61
heating costs and priorities, 29, 47
heating systems, x, 42–47
heat loss, xi, 29, 38, 74
heat pumps, 47
heat recovery ventilators (HRVs), 47
heat segregation, x, 16–17, 18
heat sinks, 9
heat tape, x, xii, 75
heat tracers, 74, 83
heat transfer, 58, 60
helical anchors, 18, 23–24
high-density polyethylene pipe (HDPE), 76
honey buckets, 80, 81, 95
hot water heating systems, 42, 44, 45–46
human waste. *see* sewage treatment systems; waste disposal systems
humidity, 49–50, 51–52
humus toilets, 89
hybrid composting-graywater treatment systems, 90–91
hydroelectric plants, 39

I
ice, as water source, 70, 72
ice creep, 22
ice-rich permafrost, 6, 15–16, 17–18
ice wedges, 4–6, 8
icing (glaciering), 69
igloos, 27
incineration of waste, 90, 92
insulations, 28–32, 47, 54, 57
Inuvik, 21, 82, 97
iron, in arctic water, 71, 76

K
Kivalina, 90
Kotzebue, 97

L
lakes, 7, 68, 69
leaks in roofs, 54
liquefied gas, 39
liquid wastes, disposal of, 96
low-emissivity (low-E) coating, 62